MITRE

Ten Strategies of a World–Class
Cybersecurity Operations Center

Carson Zimmerman

Produced by MITRE Corporate Communications and Public Affairs

International Standard Book Number: 978-0-692-24310-7

Printed in the United States of America on acid-free paper

The MITRE Corporation

202 Burlington Road • Bedford, MA 01730-1420 • (781) 271-2000

7515 Colshire Drive • McLean, VA 22102-7539 • (703) 983-6000

www.mitre.org

cyber@mitre.org

Send feedback or questions on this book to tenstrategies@mitre.org.

Acknowledgments

■··■

There are many individuals whose hard work has contributed to the creation of this book.

First of all, I would like to recognize Eric Lippart, whose many years of work in computer network defense (CND) contributed to every aspect of this book. The ten strategies outlined in the book emerged from the years we worked together to share best practices and solutions for CND across the U.S. federal government.

Some sections of this book are based, in part, on material from other MITRE work. The following sections incorporate ideas and expertise from other MITRE staff members:

- Scott Foote, Chuck Boeckman, and Rosalie McQuaid: Cyber situational awareness, Section 2.5
- Julie Connolly, Mark Davidson, Matt Richard, and Clem Skorupka: Cyber attack life cycle, Section 2.6
- Susan May: CSOC staffing, Section 7.2
- Mike Cojocea: Security information and event management (SIEM) and log management (LM) best practices, Section 8.3
- Joe Judge and Eugene Aronne: Original work on intrusion detection systems (IDS) and SIEM, Section 8.2 and Section 8.3
- Frank Posluszny: Initial concept and development of material on Cyber Threat Analysis Cells, Sections 11.1–11.6
- Kathryn Knerler: CND resources and websites, Section 11.7

- Therese Metcalf: Material on various government CSOCs, which was used throughout this book
- Bob Martin: Technical editing and glossary
- Robin Cormier: Public release and project management
- Robert Pappalardo and John Ursino: Cover design
- Susan Robertson: Book layout and diagrams

The following individuals are recognized as peer reviewers for this book: Chuck Boeckman, Mike Cojocea, Dale Johnson, Kathryn Knerler, Eric Lippart, Rick Murad, Todd O'Boyle, Lora Randolph, Marnie Salisbury, Ben Schmoker, Wes Shields, and Dave Wilburn.

This book would not have been possible without the funding and mentorship provided by MITRE's cybersecurity leadership: Gary Gagnon, Marnie Salisbury, Marion Michaud, Bill Neugent, Mindy Rudell, and Deb Bodeau. In addition, Lora Randolph's and Marnie Salisbury's advice and support were instrumental in making this book possible.

This book was also inspired by the excellent work done by the Carnegie Mellon University (CMU) Software Engineering Institute (SEI) Computer Emergency Response Team (CERT®), whose materials are referenced herein. Their copyrighted material has been used with permission. The following acknowledgement is included per CMU SEI:

> This publication incorporates portions of the "Handbook for Computer Security Incident Response Teams, 2nd Ed.," CMU/SEI-2003-HB-002, Copyright 2003 Carnegie Mellon University and "Organizational Models for Computer Security Incident Response Teams," CMU/SEI-2003-HB-001, Copyright 2003 Carnegie Mellon University with special permission from its Software Engineering Institute.
>
> Any material of Carnegie Mellon University and/or its Software Engineering Institute contained herein is furnished on an "as-is" basis. Carnegie Mellon University makes no warranties of any kind, either expressed or implied, as to any matter including, but not limited to, warranty of fitness for purpose or merchantability, exclusivity, or results obtained from use of the material. Carnegie Mellon University does not make any warranty of any kind with respect to freedom from patent, trademark, or copyright infringement. This publication has not been reviewed nor is it endorsed by Carnegie Mellon University or its Software Engineering Institute.
>
> CERT is a registered trademark of Carnegie Mellon University.

I would like to recognize the tireless efforts of the many operators, analysts, engineers, managers, and executives whose contributions to cyber defense have helped shape this book.

This book is dedicated to Kristin and Edward.

About the Cover

"Now, here, you see, it takes all the running you can do, to keep in the same place. If you want to get somewhere else, you must run at least twice as fast as that!"

— The Red Queen, to Alice, in Lewis Carroll's *Through the Looking Glass*

The adversary is constantly advancing its capabilities. Enterprise networks are always adapting to accommodate new technologies and changing business practices. The defender must expend all the effort it can just to stay in the same relative place, relative to what it must protect and defend against. Actually advancing its capabilities—matching or getting ahead of the adversary—takes that much more effort.

Stealing a concept from evolutionary biology, we draw a parallel others in cybersecurity have to the Red Queen Hypothesis. The Cybersecurity Operations Center must constantly evolve its tactics, techniques, procedures, and technologies to keep pace. This is a frequent refrain throughout the book.

Contents

Acknowledgments iii

Executive Summary 1

Introduction 3

Fundamentals 8

Strategy 1: Consolidate CND Under One Organization 44

Strategy 2: Achieve Balance Between Size and Agility 49

Strategy 3: Give the SOC the Authority to Do Its Job 71

Strategy 4: Do a Few Things Well 80

Strategy 5: Favor Staff Quality over Quantity 87

Strategy 6: Maximize the Value of Technology Purchases 108

Strategy 7: Exercise Discrimination in the Data You Gather 175

Strategy 8: Protect the SOC Mission 206

Strategy 9: Be a Sophisticated Consumer and Producer of Cyber Threat Intelligence 221

Strategy 10: Stop. Think. Respond . . . Calmly 250

References. 258

Appendix A: External Interfaces 278

Appendix B: Do We Need Our Own SOC? 282

Appendix C: How Do We Get Started? 286

Appendix D: Should the SOC Go 24x7?. 291

Appendix E: What Documents Should the SOC Maintain?. 295

Appendix F: What Should an Analyst Have Within Reach? 301

Appendix G: Characteristics of a World–Class SOC 307

Appendix H: Glossary 319

Appendix I: List of Abbreviations. 326

Index 332

Figures

Figure 1. SOC Roles and Incident Escalation. 14

Figure 2. OODA Loop. 26

Figure 3. Cyber Attack Life Cycle 30

Figure 4. Typical SOC Tool Architecture 33

Figure 5. Context to Tip-offs: Full-Spectrum CND Data 34

Figure 6. CND Tools in the Abstract. 36

Figure 7. The Four Categories of Activity. 37

Figure 8. Balancing Data Volume with Value. 38

Figure 9. Data Volume Versus Value in Practice 39

Figure 10. All Functions of CND in the SOC 45

Figure 11. Rightsizing the Constituency 50

Figure 12. Small SOC 54

Figure 13. Alternative Arrangement for a Small SOC 55

Figure 14. Example: Large SOC 57

Figure 15. Data Flows Between Central and Subordinate SOCs 68

Figure 16. Typical Career Paths Through the SOC 103

Figure 17. Basic Network Discovery Architecture 112

Figure 18. Typical IDS Architecture 120

Figure 19. IDS Signature Age Versus Usefulness in Detection 123

Figure 20. Virtualizing IDSes 135

Figure 21. SIEM Overview 155

Figure 22. SIEM: Supporting the Event Life Cycle from Cradle to Grave 157

Figure 23. SIEM Value and SOC Staffing Versus Maturity 158

Figure 24. Log Data Delivery Options and SIEM Tiering 160

Figure 25. Overlap Between SIEM, Network Management System, and LM. 163

Figure 26. Instrumenting an Internet Gateway .182

Figure 27. Instrumenting an External-Facing DMZ 184

Figure 28. Copying Network Traffic 208

Figure 29. Data Diode 212

Figure 30. Protected SOC Architecture 217

Figure 31. CTAC and SOC Relationships to Other Organizations 229

Figure 32. People, Process, Technology CND Enablers 287

Tables

Table 1. SOC Capabilities 19

Table 2. Cyber Attack Life Cycle 31

Table 3. Ops Tempo of the Attacker Versus the Defender 41

Table 4. SOC Templates 51

Table 5. Considerations for SOC Placement. 59

Table 6. Differences in Roles for Tiered Approach 67

Table 7. Service Templates 82

Table 8. Automated Versus Manual Network Mapping 113

Table 9. Advantages and Disadvantages of Intrusion Detection Elements 121

Table 10. Commercial NIDS/NIPS Characteristics 133

Table 11. COTS Versus FOSS for Network Monitoring 137

Table 12. SIEM and Log Aggregation Systems Compared. 162

Table 13. Best-of-Breed SIEM Characteristics .165

Table 14. Network Sensor Placement Considerations 178

Table 15. Host Monitoring Placement Considerations 181

Table 16. Comparison of Common Security-Relevant Data Sources 194

Table 17. Approaches to Tuning Data Sources. .201

Table 18. Suggested Data Retention Time Frames 205

Table 19. Traffic Redirection Options 210

Table 20. Sharing Sensitive Information 219

Table 21. CTAC Artifacts 225

Table 22. Other CTAC Work Products. 226

Table 23. CTAC Relationship to SOC Elements 227

Table 24. Case Tracking Approaches 255

Table 25. SOC Touch Points 278

Table 26. Scoring the Need for a SOC 283

Table 27. Example #1: Big Toy Manufacturing, Inc. 284

Table 28. Example #2: Big Government Agency 285

Table 29. Document Library. 295

Table 30. Analyst Resources 301

Table 31. Tier 2 Tools 305

Executive Summary

Today's cybersecurity operations center (CSOC) should have everything it needs to mount a competent defense of the ever-changing information technology (IT) enterprise. This includes a vast array of sophisticated detection and prevention technologies, a virtual sea of cyber intelligence reporting, and access to a rapidly expanding workforce of talented IT professionals. Yet, most CSOCs continue to fall short in keeping the adversary—even the unsophisticated one—out of the enterprise.

The deck is clearly stacked against the defenders. While the adversary must discover only one way in, the defenders must defend all ways in, limit and assess damage, and find and remove adversary points of presence in enterprise systems. And cybersecurity experts increasingly recognize that sophisticated adversaries can and will establish lasting footholds in enterprise systems. If this situation were not bad enough, more often than not, we are our own worst enemy. Many CSOCs expend more energy battling politics and personnel issues than they do identifying and responding to cyber attacks. All too often, CSOCs are set up and operate with a focus on technology, without adequately addressing people and process issues. The main premise of this book is that a more balanced approach would be more effective.

This book describes the ten strategies of effective CSOCs—regardless of their size, offered capabilities, or type of constituency served. The strategies are:

1. Consolidate functions of incident monitoring, detection, response, coordination, and computer network defense tool engineering, operation, and maintenance under one organization: the CSOC.

2. Achieve balance between size and visibility/agility, so that the CSOC can execute its mission effectively.

3. Give the CSOC the authority to do its job through effective organizational placement and appropriate policies and procedures.

4. Focus on a few activities that the CSOC practices well and avoid the ones it cannot or should not do.

5. Favor staff quality over quantity, employing professionals who are passionate about their jobs, provide a balance of soft and hard skills, and pursue opportunities for growth.

6. Realize the full potential of each technology through careful investment and keen awareness of—and compensation for—each tool's limitations.

7. Exercise great care in the placement of sensors and collection of data, maximizing signal and minimizing noise.

8. Carefully protect CSOC systems, infrastructure, and data while providing transparency and effective communication with constituents.

9. Be a sophisticated consumer and producer of cyber threat intelligence, by creating and trading in cyber threat reporting, incident tips and signatures with other CSOCs.

10. Respond to incidents in a calm, calculated, and professional manner.

In this book, we describe each strategy in detail, including how they crosscut elements of people, process, and technology. We deeply explore specific areas of concern for CSOCs, ranging from how many analysts a CSOC needs to where to place sensor technologies.

Chapter 1

Introduction

1.1 Background

Ensuring the confidentiality, integrity, and availability of the modern information technology (IT) enterprise is a big job. It incorporates many tasks, from robust systems engineering and configuration management (CM) to effective cybersecurity or information assurance (IA) policy and comprehensive workforce training. It must also include cybersecurity operations, where a group of people are charged with monitoring and defending the enterprise against all measures of cyber attack. The focal point for security operations and computer network defense (CND) in the large enterprise is the cybersecurity operations center (CSOC, or simply SOC). Virtually all large IT enterprises have some form of SOC, and their importance is growing as "cyber" becomes increasingly important to the mission.

The MITRE Corporation supports a number of U.S. government SOCs, which go by many names: Computer Security Incident Response Team (CSIRT), Computer Incident Response Team (CIRT), Computer Security Incident Response Capability (CSIRC), Network Operations and Security Center (NOSC), and, of course, CSOC. As a corporation that operates several federally funded research and development centers, MITRE's support to these entities spans many years, with staff members who support the full scope of the SOC mission, from standing up new SOCs to enhancing existing CND capabilities. Operational activities range from analyzing network traffic captures to editing incident escalation procedures and architecting enterprise sensor grids. We draw upon these experiences in developing this book.

Since many SOCs were first stood up in the 1990s, several movements in IT have changed the way the CND mission is executed:

1. The rise of the advanced persistent threat (APT) [1] and an acceleration in the evolution of the adversary's tactics, techniques, and procedures (TTPs)
2. A movement toward IT consolidation and cloud-based computing
3. The exponential growth in mobile devices, obscuring where enterprise borders truly lie
4. A transition from network-based buffer overflow attacks to client-side attacks.

Despite the fact that the practice of CND and incident response is more than 20 years old, SOCs still wrangle with fundamental issues related not just to the technologies they must work with day to day but with larger issues related to people and process, from how to handle incident escalation to where the SOC belongs on the constituency organization chart.

The majority of recognized materials about SOCs were published between 1998 and 2005 [2], [3], [4], [5], [6], [7], [8], [9], [10], [11], [12], [13]. Recent publications are somewhat more focused in scope [14], [15], cover mostly tools [16], or discuss cyber warfare doctrine [17], [18]. With some exceptions [19], [20], comparatively few comprehensive works on CND have been written in the last ten years. Furthermore, most references about CND focus on technology to the exclusion of people and process or cover people and process at length without bringing in the elements of technology and tools.

There have been significant changes in cyber threat and technology but few updates on the fundamental "hows" and "whys" of CND. Moreover, most general references on incident response and intrusion detection are tailored toward small- to medium-sized enterprises and focus on technology. Far less material on enabling security operations in large enterprises exists. People and process issues are increasingly the primary impediment to effective CND.

Our firsthand experience, along with these trends and observations, motivated us to write and publish a book that addresses the elements of a modern, world-class SOC.

1.2 Purpose

This book aims to help those who have a role in cybersecurity operations to enhance their CND capability. To do so, we leveraged observations and proven approaches across of all three CND elements (people, process, and technology). The structure of this book directly addresses ten strategies that, when followed, lead to a vastly enhanced CND capability. We address common questions often pondered by SOCs—new and old, big and small—in the context of their enterprise and mission space, leveraging recent advancements in security operations and incident response best practices.

This book's objectives are to help organizations:

1. Articulate a coherent message of "this is the way CND is done"
2. Translate the lessons learned from leading, mature SOCs to new and struggling SOCs
3. Provide context and options for critical SOC architecture, tools, and process decisions
4. Demonstrate the way CND has evolved in the face of APTs
5. Differentiate the roles of different SOCs, given various constituency sizes and missions
6. Maximize the value of SOC staff and technology investments.

1.3 Audience

If you are part of, support, frequently work with, manage, or are trying to stand up a SOC, this book is for you. Its primary audience is SOC managers, technical leads, engineers, and analysts. Portions of this book can be used also as a reference by those who interface with SOCs on a weekly or daily basis to better understand and support CND operations. These include chief information security officers (CISOs); network operations center (NOC) personnel; senior IT system administrators (sysadmins); counterintelligence and law enforcement personnel who work cyber cases; and information systems security engineers (ISSEs), officers (ISSOs), and managers (ISSMs).

Anyone reading this book is assumed to have a general understanding of IT and security concepts and a general awareness of cyber threats. A specific background in computer science, engineering, networking, or system administration is especially helpful.

If your IT enterprise has fewer than 1,000 users or nodes, your resourcing may not be sufficient to support a SOC. In this case, outsourcing major elements of CND may be worth considering. Therefore, while parts of this book may be informative, please remember that many of the considerations and assumptions are made with medium and large enterprises in mind.

Conversely, if your IT enterprise measures in the many thousands of hosts/Internet Protocol (IP) addresses or users, you are likely best served by a consolidated CND capability—meaning that this book is for you. In Section 4, we also address approaches to enterprises measuring in the millions of nodes or users, such as those spanning multiple government agencies.

We had U.S. executive branch agencies in mind when we wrote this book, including federal civilian agencies, the Department of Defense (DoD), and the Intelligence

Community. Some material is most applicable when applied to this space, and assumptions unique to the U.S. federal executive branch influence some suggestions. However, the vast majority of the best practices approaches discussed herein are applicable to other areas of government (United States or otherwise). They also apply to the commercial sector, non-profits, and academia. This book therefore does not focus on specific mandates and guidance promulgated by the National Institute of Standards and Technology (NIST), DoD, the Director of National Intelligence, or the Office of Management and Budget. However, when mandates and guidance are relevant to the organization and operations of a SOC, they are briefly discussed.

1.4 How to Use This Book

This book has been written as if you, the reader, have been given the task of operating a SOC. We take this approach to emphasize the operational and practical "real-world" nature of defending a network. Regardless of your background in CND, the book is intended to convey advice and best practices culled from a number of SOCs that MITRE has supported over the years. As a result, this book adopts two key conventions. First, tangible, concrete guidance is provided wherever possible, favoring detailed recommendations and advice that work 90 percent of the time. Second, many other excellent references on some of the technological focus areas of CND exist, and this book leverages those references wherever possible, while focusing on the integration of people, process, and technology.

Throughout the book, key points that we want to draw your attention to will be designated as such:

> **This is a really important point, worthy of your consideration.**

The book is organized as follows:
1. **Section 1** introduces the book's subject, purpose, scope, and approach.
2. **Section 2** discusses the fundamentals of SOCs. It targets readers who do not have a strong background in CND. It includes a SOC's basic mission, capabilities, and technologies. Section 2.3 and Section 2.4 are referenced frequently throughout the remainder of the book, so those familiar with CND may want first to review these sections before proceeding.
3. **Sections 3 through 12** describe the ten strategies of an effective SOC. These sections take the fundamentals covered in Section 2 and discuss them in practice. They break down key design, architecture, and procedural issues in detail. These

sections house the main body of material covered in this book and crosscut issues of people, process, and technology.

4. The **Appendices** address issues not covered in the main body. These are relevant to supporting a world-class CND capability but do not fit within the structure of the ten strategies. This section also contains a Glossary and List of Abbreviations.

1.5 Scope

This book integrates CND's people, process, and technology elements. Its style is not only descriptive about the fundamentals of CND but prescriptive about how best to accomplish the CND mission. We take advantage of excellent existing materials that cover purely technical or purely people/process aspects of CND. As a result, we attempt to avoid the following topics:

1. Bits and bytes of transmission control protocol (TCP)/IP and packet analysis [21], [22], [23]
2. Media and system forensics [24], [25], [26], [27], [28], [29]
3. Legal aspects of network monitoring, such as privacy laws, details on chain of custody, and specific legal or regulatory requirements for retention of audit data
4. Vulnerability assessment and penetration testing [30], [31], [32], [33], [34], [35]
5. Malware analysis and exploits [36], [37], [38], [39], [40]
6. How to use specific CND technologies such as intrusion detection systems (IDSes) or security information and event management (SIEM)
7. Point-in-time comparisons of specific CND technologies and point products [41]
8. Detailed description of how various cyber attacks work
9. Issues that are out of the scope of CND:
   a. Network and telecommunications monitoring and operations
   b. Physical security operations (e.g., "gates, guards, guns")
10. Compliance with specific laws and regulations.

While many references are available, we especially use material from the SysAdmin, Audit, Networking, and Security (SANS) Institute and Carnegie Mellon Software Engineering Institute (SEI) Computer Emergency Response Team (CERT) because of their high quality, acceptance across industry, and open availability.

Chapter 2
Fundamentals

....

In this section, we cover the fundamental concepts of a SOC: what a SOC is, what it does, the major tools it uses to accomplish its mission, and its key mission drivers. Much of this material resurfaces in later sections.

Even if you have been involved with incident response or security operations in the past, it may be helpful to review this section because it establishes key terminology and drivers used throughout the book.

2.1 What Is a SOC?

A SOC is defined primarily by what it does—CND. We adapted the definition from [42], characterizing CND as:

> The practice of defense against unauthorized activity within computer networks, including monitoring, detection, analysis (such as trend and pattern analysis), and response and restoration activities.

There are many terms that have been used to reference a team of cybersecurity experts assembled to perform CND. They include:

- Computer Security Incident Response Team (CSIRT)
- Computer Incident Response Team (CIRT)
- Computer Incident Response Center (or Capability) (CIRC)
- Computer Security Incident Response Center (or Capability) (CSIRC)
- Security Operations Center (SOC)
- Cybersecurity Operations Center (CSOC)

- Computer Emergency Response Team [4, pp. 10–11] (CERT[1]).

Common variations in the name include words such as "network," "computer," "security," "cyber," "information technology," "emergency," "incident," "operations," or "enterprise."

The acronym "CSIRT" is the most technically accurate term that may be used in reference to the team of personnel assembled to find and respond to intrusions. However, its usage is far from universal, most CSIRTs go by some designation other than "CSIRT," and its usage has waned in recent years. As a result, identifying them by name alone is not always easy. Many (if not most) cybersecurity professionals use "SOC" colloquially to refer to a CSIRT, even though using the term "SOC" in such a manner is not entirely correct.[2]

Purity of vocabulary might serve as a distraction to many readers, and we wish to maintain consistency with modern common usage. As a result, in this book, we use "SOC."

We combine definitions of CSIRT from [42]and [43] to define "SOC:"

> **A SOC is a team primarily composed of security analysts organized to detect, analyze, respond to, report on, and prevent cybersecurity incidents.**

A SOC provides services to a set of customers referred to as a **constituency**—a bounded set of users, sites, IT assets, networks, and organizations. Combining definitions from [43]and [8], j87a constituency can be established according to organizational, geographical, political, technical, or contractual demarcations.

Once again borrowing from the historical definition of "CSIRT," [44]articulates three criteria that an organization must meet to be considered a CSIRT, which we hold over in applying to a SOC. In order for an organization to be considered a SOC, it must:

1. Provide a means for constituents to report suspected cybersecurity incidents
2. Provide incident handling assistance to constituents
3. Disseminate incident-related information to constituents and external parties.

The SOC provides a set of services to the constituency that is related to the core mission of incident detection and response, such as security awareness training or vulnerability assessment. Compare a SOC's services to its constituency to the way a fire or paramedic station operates [3, p. 11]. Firefighters' and other first responders' primary role is to help

---

1 Computer Emergency Response Team (CERT) is registered in the U.S. Patent and Trademark Office by Carnegie Mellon University.
2 It should also be noted that some *physical* security operations centers also go by "SOC." Thus, in the title of this book, we used "Cyber" to disambiguate our topic.

people in emergencies, but a lot more goes into the mission of protecting the public from harm. It includes, among other activities, fire safety education, home and business fire inspections, and first-responder training. Outreach goes a long way toward preventing fires—and preventing security incidents.

Some fire departments have the resources to perform a detailed post-incident analysis of why a fire started and how it spread. Similarly, some SOCs have the skills and resources to perform detailed forensics on compromised systems. Others, however, must call on partner SOCs or external resources when in-depth forensics must be performed. Section 2.4 discusses the range of services that a SOC may provide.

2.2 Mission and Operations Tempo

SOCs can range from small, five-person operations to large, national coordination centers. A typical midsize SOC's mission statement typically includes the following elements:

1. Prevention of cybersecurity incidents through proactive:
    a. Continuous threat analysis
    b. Network and host scanning for vulnerabilities
    c. Countermeasure deployment coordination
    d. Security policy and architecture consulting.
2. Monitoring, detection, and analysis of potential intrusions in real time and through historical trending on security-relevant data sources
3. Response to confirmed incidents, by coordinating resources and directing use of timely and appropriate countermeasures
4. Providing situational awareness and reporting on cybersecurity status, incidents, and trends in adversary behavior to appropriate organizations
5. Engineering and operating CND technologies such as IDSes and data collection/analysis systems.

Of these responsibilities, perhaps the most time-consuming is the consumption and analysis of copious amounts of security-relevant data. Among the many security-relevant data feeds a SOC is likely to ingest, the most prominent are often IDSes. IDSes are systems placed on either the host or the network to detect potentially malicious or unwanted activity that warrants further attention by the SOC analyst. Combined with security audit logs and other data feeds, a typical SOC will collect, analyze, and store tens or hundreds of millions of security events every day.

According to [42], an **event** is "Any observable occurrence in a system and/or network. Events sometimes provide indication that an incident is occurring" (e.g., an **alert** generated by an IDS or a security audit service). An event is nothing more than raw data. It takes human **analysis**—the process of evaluating the meaning of a collection of security-relevant data, typically with the assistance of specialized tools—to establish whether further action is warranted.

Handling the constant influx of data is analogous to triaging patients in an emergency situation—there are hundreds of wounded people, so who deserves attention first? Adapting the definition found in [4, p. 16], **triage** is the process of sorting, categorizing, and prioritizing incoming events and other requests for SOC resources.

A SOC typically will designate a set of individuals devoted to real-time triage of alerts, as well as fielding phone calls from users and other routine tasks. This group is often referred to as **Tier 1**.[1] If Tier 1 determines that an alert reaches some predefined threshold, a **case** is created and **escalated** to Tier 2. This threshold can be defined according to various types of potential "badness" (type of incident, targeted asset or information, impacted mission, etc.). Usually, the time span Tier 1 has to examine each **event of interest** is between one and 15 minutes. It depends on the SOC's escalation policy, concept of operations (CONOPS), number of analysts, size of constituency, and event volume. Tier 1 members are discouraged from performing in-depth analysis, as they must not miss events that come across their real-time consoles. If an event takes longer than several minutes to evaluate, it is escalated to Tier 2.

Tier 2 accepts cases from Tier 1 and performs in-depth analysis to determine what actually happened—to the extent possible, given available time and data—and whether further action is necessary. Before this decision is made, it may take weeks to collect and inspect all the necessary data to determine the event's extent and severity. Because Tier 2 is not responsible for real-time monitoring and is staffed with more experienced analysts, it is able to take the time to fully analyze each activity set, gather additional information, and coordinate with constituents. It is generally the responsibility of Tier 2 (or above) to determine whether a potential incident occurred. According to [42], an **incident** is:

> . . . an assessed occurrence that actually or potentially jeopardizes the confidentiality, integrity, or availability of an information system; or the information the system processes, stores, or transmits; or that constitutes a violation or imminent threat of violation of security policies, security procedures or acceptable use policies.

---

1 We follow a hierarchy for analysis and escalation within the SOC that originates from tiering within an IT help desk and system administration organizations. This arrangement should not be confused with *tiered SOCs,* whereby multiple SOCs are organized within one large constituency (which is discussed in Section 4.3.7).

A single event can spawn an incident, but, for every incident, there are likely thousands or millions of events that are simply benign.

Some SOCs also will allocate resources from one or more of their own sections (sometimes including Tier 2) to look for all the unstructured indicators of incidents that are outside the scope of Tier 1's duties. Indeed, many cases stem from the non-routine indicators and analytics that Tier 1 does not have the capability to work with but that Tier 2 does. In larger SOCs, these teams work in concert to find and evaluate the disposition of suspicious or anomalous activity on constituency networks.

The SOC typically will leverage internal and external resources in **response** to and recovery from the incident. It is important to recognize that a SOC may not always deploy countermeasures at the first sign of an intrusion. There are three reasons for this:

1. The SOC wants to be sure that it is not blocking benign activity.
2. A response action could impact a constituency's mission services more than the incident itself.
3. Understanding the extent and severity of the intrusion by watching the adversary is sometimes more effective than performing static forensic analysis on compromised systems, once the adversary is no longer present.

To determine the nature of the attack, the SOC often must perform advanced **forensic analysis** on artifacts such as hard drive images or full-session packet capture (PCAP), or **malware reverse engineering** on malware samples collected in support of an incident. Sometimes, forensic evidence must be collected and analyzed in a legally sound manner. In such cases, the SOC must observe greater rigor and repeatability in its procedures than would otherwise be necessary.

When the signs of an attack are understood well enough to encode a computer-readable IDS signature, the attack may be prevented in-line, as with a host intrusion prevention system (HIPS) or network intrusion prevention system (NIPS). While such systems typically are used to prevent the most basic attacks, the extent to which they can automate analysis is limited. Human analysis is always needed to run a major incident to ground. A number of technologies enable the SOC to comb through millions of events every day, supporting the incident life cycle from cradle to grave. SIEM tools collect, store, correlate, and display myriad security-relevant data feeds, supporting triage, analysis, escalation, and response activities. IDS and intrusion prevention systems (IPSs) are two of many systems that feed SIEM. The current perspective on IDS/IPS technologies suggests that they may be necessary but that they are certainly not sufficient in incident detection (though they are still relevant and useful—this issue is discussed at length in Section 8.2). The variety of data feeds together support both the tip-off and contextual "ground truth" needed to support incident analysis. Any one source of security events has little value; the whole is greater than the sum of the parts.

The SOC does not just consume data from its constituency; it also folds in information from a variety of external sources that provides insight into threats, vulnerabilities, and adversary **TTPs**. This information is called cyber intelligence (intel), and it includes cyber news feeds, signature updates, incident reports, threat briefs, and vulnerability alerts. As the defender, the SOC is in a constant arms race to maintain parity with the changing environment and threat landscape. Continually feeding cyber intel into SOC monitoring tools is key to keeping up with the threat. In a given week, the SOC likely will process dozens of pieces of cyber intel that can drive anything from IDS signature updates to emergency patch pushes. A SOC must discriminate among the cyber intel that it harvests; intel must be timely, relevant, accurate, specific, and actionable about the incident, vulnerability, or threat it describes.

In describing all of the roles the different parts of the SOC carry out, recognize that each SOC will divide roles differently and that overlap can exist between functions. Some SOCs perform in-depth analysis and response in a single Tier 2 section. Others will break some of these functions into a third tier. Rather than covering all of the permutations, the general escalation path and incident response roles of a SOC are illustrated in Figure 1.

Other entities in the constituency are similar to SOCs, but they should be recognized as distinct and separate. A SOC is distinct from:

- A NOC because a SOC is primarily looking for cyber attacks, whereas a NOC is concerned with operating and maintaining network equipment
- A chief information officer (CIO) or CISO office because the SOC is a real-time operational capability, and its monitoring efforts are not usually intended to support strategic planning (though some SOCs may report directly to a CISO or CIO)
- An Information Security Continuous Monitoring (ISCM) program [45] because the SOC is responsible for incident detection and response
- An ISSO or ISSM shop because the SOC is responsible for monitoring and responding to the full-scope cyber threat across the entire constituency, whereas ISSOs are usually involved with IT compliance and ensuring the security of specific systems
- Computer network exploitation (CNE) or computer network attack (CNA) teams because SOCs do not normally perform attack or penetration activities outside their constituency
- Physical security monitoring (e.g., "gates, guards, and guns") because a SOC is concerned with the cyber domain, whereas physical security monitoring is primarily concerned with protecting physical assets and ensuring personnel safety

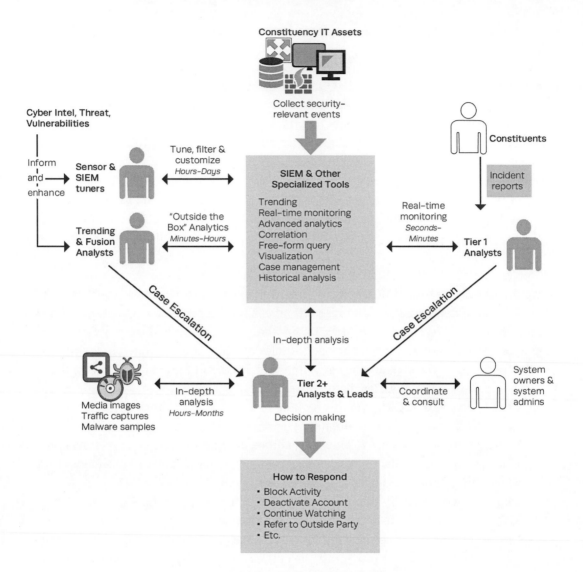

Figure 1. SOC Roles and Incident Escalation

- Law enforcement because SOCs are usually not investigative bodies; while they may find intrusions that result in legal action, their primary duty is usually not the collection, analysis, and presentation of evidence that will be used in legal proceedings.

These delineations highlight the fact that SOCs must work with all of these groups on a regular basis, and they must maintain skill sets in many areas of IT and cybersecurity.

2.3 Characteristics

Because SOCs can vary so widely in size and structure, key qualities must be identified to describe these differences. We begin with three characteristics of the SOC before covering its various capabilities in detail. These characteristics are:

1. **Organizational relationship** to the SOC's constituency
2. The **distribution of resources** that comprise the SOC (e.g., its **organizational model**)
3. The **authority** it exercises over its constituency.

We base the discussion on material from Carnegie Mellon CERT [4] [8], with updates or conceptual changes noted where applicable. These characteristics will be reused throughout the remainder of the book.

2.3.1 Constituency Membership

SOCs are either part of the organization they serve or external to it.

SOCs that are **external** to the constituency include managed security service providers (MSSPs) run by a corporation providing services to paying clients. They also include SOCs that are product focused, such as those run by vendors, which must rapidly respond to security vulnerabilities in products that serve a large customer base.

In most cases, a SOC is part of the organization it defends; therefore, its relationship is considered **internal**. An internal SOC's constituency is composed of the users and IT assets that belong to one autonomous organization, such as a government agency, university, or corporation, of which the SOC is also a member. In these cases, it is also common to find a chief executive officer (CEO) and CIO with cognizance of the entire organization. Finally, this constituency is most likely subdivided into multiple business units, each located at one or more disparate geographic locations.

> **Much of the advice given in this book has internal constituency membership in mind.**

This definition is consistent with consistency membership defined in [4, p. 12].

2.3.2 Organizational Model

We categorize SOCs that are internal to the constituency into five organizational models of how the team is comprised, following that of Section 3.2 in [4]:

1. **Security team.** No standing incident detection or response capability exists. In the event of a computer security incident, resources are gathered (usually from within the constituency) to deal with the problem, reconstitute systems, and then

stand down. Results can vary widely as there is no central watch or consistent pool of expertise, and processes for incident handling are usually poorly defined. Constituencies composed of fewer than 1,000 users or IPs usually fall into this category.

2. **Internal distributed SOC.** A standing SOC exists but is primarily composed of individuals whose organizational position is outside the SOC and whose primary job is IT or security related but not necessarily CND related. One person or a small group is responsible for coordinating security operations, but the heavy lifting is carried out by individuals who are matrixed in from other organizations. SOCs supporting a small- to medium-sized constituency, perhaps 500 to 5,000 users or IPs, often fall into this category.

3. **Internal centralized SOC.** A dedicated team of IT and cybersecurity professionals comprise a standing CND capability, providing ongoing services. The resources and the authorities necessary to sustain the day-to-day network defense mission exist in a formally recognized entity, usually with its own budget. This team reports to a SOC manager who is responsible for overseeing the CND program for the constituency. Most SOCs fall into this category, typically serving constituencies ranging from 5,000 to 100,000 users or IP addresses.

4. **Internal combined distributed and centralized SOC.** The SOC is composed of both a central team (as with internal centralized SOCs) and resources from elsewhere in the constituency (as with internal distributed SOCs). Individuals supporting CND operations outside of the main SOC are *not* recognized as a separate and distinct SOC entity. For larger constituencies, this model strikes a balance between having a coherent, synchronized team and maintaining an understanding of edge IT assets and enclaves. SOCs with constituencies in the 25,000–500,000 user/IP range may pursue this approach, especially if their constituency is geographically distributed or they serve a highly heterogeneous computing environment.

5. **Coordinating SOC.** The SOC mediates and facilitates CND activities between multiple subordinate distinct SOCs, typically for a large constituency, perhaps measured in the millions of users or IP addresses. A coordinating SOC usually provides consulting services to a constituency that can be quite diverse. It typically does not have active or comprehensive visibility down to the end host and most often has limited authority over its constituency. Coordinating SOCs often serve as distribution hubs for cyber intel, best practices, and training. They also can offer analysis and forensics services, when requested by subordinate SOCs.

In practice, few SOCs fall neatly within just one of these categories—most SOCs fall on a spectrum somewhere between purely distributed and purely centralized. In addition, many coordinating SOCs will harmonize the activities of subordinate SOCs and will also perform their own direct monitoring and response operations. In any event, a SOC must balance a deep understanding of CND topics with knowledge down to the end asset and mission. Together, these form a SOC's "street cred" (i.e., constituents' and other SOCs' perception of its expertise and the value it adds). Since this is absolutely vital to a SOC's success, we return to this theme many times throughout the book. Section 4 discusses sizing SOC models to the constituency in which they serve.

2.3.3 Authority

Authority describes the amount of discretion the SOC has in directing actions that impact constituency assets, with or without permission from, or coordination with, other groups. We leverage the authority framework described in [4, p. 37], with some modifications. A SOC can exert three levels of authority:

1. **No authority.** A SOC can suggest or influence constituents regarding actions they should take. However, the SOC has neither the formal means to exert pressure nor a superior organizational element willing or able to do so. It is entirely up to constituents to heed or ignore the SOC's recommendations.

2. **Shared authority.** A SOC can make recommendations to constituency executives (e.g., CIOs, CISOs, CEOs, system owners) who have various authorities to enact change. These recommendations are weighed against input from other stakeholders before a decision is made, giving the SOC a vote but not the final say.

3. **Full authority.** A SOC can direct constituents to take certain actions, without seeking or waiting for the approval or support from any higher level party. The SOC's ability to dictate changes is recognized at least in practice, if not formally.

A SOC's authority is never absolute, even if it has "full" authority within some scope of its operations. A SOC's formal authorities can be applied up to a point, beyond which the SOC must turn to influence rather than mandate. For aggressive countermeasures or response such as disabling a key corporate server, high-level agreement and understanding is needed.

While a SOC may have the ability to unplug a system from the network because something bad *did happen*, it usually may not mandate a configuration change for the enterprise because something bad *might happen*. In fact, some SOCs' authorities are written so that they may only leverage the bulk of their authority when they declare a serious incident. In practice, all SOCs leverage both formal and perceived internal authorities, as well as the influence and authorities of their seniors.

Diverging from [4], we classify authority in two stages of the incident life cycle:

1. **Reactive.** Responsive measures taken after an incident is either suspected or confirmed. Actions are usually more tactical in nature—they are temporary and impact only those constituents and systems that are directly involved in an incident. Examples include logical or physical isolation of a host, log collection from a server, or an ad hoc collection of artifacts.

2. **Proactive.** Measures taken in preemption of a perceived threat, before direct evidence of an incident is uncovered. These actions are more strategic in nature—they are usually durable and impact substantial portions of the constituency. Examples include an emergency push of patches to an entire Windows domain, domain name system (DNS) black holes or IP blocks for networks belonging to an adversary, an out-of-cycle password reset for users of a Web-facing portal, or a change to a security configuration policy.

Regardless of the authorities vested in the SOC, nothing happens without a close working relationship with system owners, sysadmins, and IT executives. Section 5 discusses SOC authorities in practice, and Appendix A lists the SOC's many touch points.

2.4 Capabilities

A SOC satisfies the constituency's network monitoring and defense needs by offering a set of **services**. We recognize the list of SOC services from [8]. However, since its publication in 2003, SOCs have matured and adapted to increased demands, a changing threat environment, and tools that have dramatically enhanced the state of the art in CND operations. We also wish to articulate the full scope of what a SOC may do, regardless of whether a particular function serves the constituency, the SOC proper, or both. As a result, this book subsumes SOC services into a comprehensive list of SOC **capabilities**.

Table 1 is a comprehensive list of common capabilities that a SOC may provide, with the understanding that any one SOC will *never* cover all of these. As described in Section 6, the SOC's management chain is responsible for picking and choosing what capabilities best fits its constituency's needs, given political and resource constraints.

Table 1. SOC Capabilities

Name	Description
Real–Time Analysis	
Call Center	Tips, incident reports, and requests for CND services from constituents received via phone, email, SOC website postings, or other methods. This is roughly analogous to a traditional IT help desk, except that it is CND specific.
Real–Time Monitoring and Triage	Triage and short-turn analysis of real-time data feeds (such as system logs and alerts) for potential intrusions. After a specified time threshold, suspected incidents are escalated to an incident analysis and response team for further study. Usually synonymous with a SOC's Tier 1 analysts, focusing on real-time feeds of events and other data visualizations. Note: This is one of the most easily recognizable and visible capabilities offered by a SOC, but it is meaningless without a corresponding incident analysis and response capability, discussed below.
Intel and Trending	
Cyber Intel Collection and Analysis	Collection, consumption, and analysis of cyber intelligence reports, cyber intrusion reports, and news related to information security, covering new threats, vulnerabilities, products, and research. Materials are inspected for information requiring a response from the SOC or distribution to the constituency. Intel can be culled from coordinating SOCs, vendors, news media websites, online forums, and email distribution lists.
Cyber Intel Distribution	Synthesis, summarization, and redistribution of cyber intelligence reports, cyber intrusion reports, and news related to information security to members of the constituency on either a routine basis (such as a weekly or monthly cyber newsletter) or a non-routine basis (such as an emergency patch notice or phishing campaign alert).
Cyber Intel Creation	Primary authorship of new cyber intelligence reporting, such as threat notices or highlights, based on primary research performed by the SOC. For example, analysis of a new threat or vulnerability not previously seen elsewhere. This is usually driven by the SOC's own incidents, forensic analysis, malware analysis, and adversary engagements.
Cyber Intel Fusion	Extracting data from cyber intel and synthesizing it into new signatures, content, and understanding of adversary TTPs, thereby evolving monitoring operations (e.g., new signatures or SIEM content).
Trending	Long-term analysis of event feeds, collected malware, and incident data for evidence of malicious or anomalous activity or to better understand the constituency or adversary TTPs. This may include unstructured, open-ended, deep-dive analysis on various data feeds, trending and correlation over weeks or months of log data, "low and slow" data analysis, and esoteric anomaly detection methods.
Threat Assessment	Holistic estimation of threats posed by various actors against the constituency, its enclaves, or lines of business, *within the cyber realm*. This will include leveraging existing resources such as cyber intel feeds and trending, along with the enterprise's architecture and vulnerability status. Often performed in coordination with other cybersecurity stakeholders.

Table 1. SOC Capabilities

Name	Description
Incident Analysis and Response	
Incident Analysis	Prolonged, in-depth analysis of potential intrusions and of tips forwarded from other SOC members. This capability is usually performed by analysts in tiers 2 and above within the SOC's incident escalation process. It must be completed in a specific time span so as to support a relevant and effective response. This capability will usually involve analysis leveraging various data artifacts to determine the who, what, when, where, and why of an intrusion—its extent, how to limit damage, and how to recover. An analyst will document the details of this analysis, usually with a recommendation for further action.
Tradecraft Analysis	Carefully coordinated adversary engagements, whereby SOC members perform a sustained "down-in-the-weeds" study and analysis of adversary TTPs, in an effort to better understand them and inform ongoing monitoring. This activity is distinct from other capabilities because (1) it sometimes involves ad hoc instrumentation of networks and systems to focus on an activity of interest, such as a honeypot, and (2) an adversary will be allowed to continue its activity without immediately being cut off completely. This capability is closely supported by trending and malware and implant analysis and, in turn, can support cyber intel creation.
Incident Response Coordination	Work with affected constituents to gather further information about an incident, understand its significance, and assess mission impact. More important, this function includes coordinating response actions and incident reporting. This service does not involve the SOC directly implementing countermeasures.
Countermeasure Implementation	Actual implementation of response actions to an incident to deter, block, or cut off adversary presence or damage. Possible countermeasures include logical or physical isolation of involved systems, firewall blocks, DNS black holes, IP blocks, patch deployment, and account deactivation.
On-site Incident Response	Work with constituents to respond and recover from an incident on-site. This will usually require SOC members who are already located at, or who travel to, the constituent location to apply hands-on expertise in analyzing damage, eradicating changes left by an adversary, and recovering systems to a known good state. This work is done in partnership with system owners and sysadmins.
Remote Incident Response	Work with constituents to recover from an incident remotely. This involves the same work as on-site incident response. However, SOC members have comparatively less hands-on involvement in gathering artifacts or recovering systems. Remote support will usually be done via phone and email or, in rarer cases, remote terminal or administrative interfaces such as Microsoft Terminal Services or Secure Shell (SSH).
Artifact Analysis	
Forensic Artifact Handling	Gathering and storing forensic artifacts (such as hard drives or removable media) related to an incident in a manner that supports its use in legal proceedings. Depending on jurisdiction, this may involve handling media while documenting chain of custody, ensuring secure storage, and supporting verifiable bit-by-bit copies of evidence.

Table 1. SOC Capabilities

Name	Description
Malware and Implant Analysis	Also known as malware reverse engineering or simply "reversing." Extracting malware (viruses, Trojans, implants, droppers, etc.) from network traffic or media images and analyzing them to determine their nature. SOC members will typically look for initial infection vector, behavior, and, potentially, informal attribution to determine the extent of an intrusion and to support timely response. This may include either static code analysis through decompilation or runtime/execution analysis (e.g., "detonation") or both. This capability is primarily meant to support effective monitoring and response. Although it leverages some of the same techniques as traditional "forensics," it is not necessarily executed to support legal prosecution.
Forensic Artifact Analysis	Analysis of digital artifacts (media, network traffic, mobile devices) to determine the full extent and ground truth of an incident, usually by establishing a detailed timeline of events. This leverages techniques similar to some aspects of malware and implant analysis but follows a more exhaustive, documented process. This is often performed using processes and procedures such that its findings can support legal action against those who may be implicated in an incident.
SOC Tool Life–Cycle Support	
Border Protection Device O&M	Operation and maintenance (O&M) of border protection devices (e.g., firewalls, Web proxies, email proxies, and content filters). Includes updates and CM of device policies, sometimes in response to a threat or incident. This activity is closely coordinated with a NOC.
SOC Infrastructure O&M	O&M of SOC technologies outside the scope of sensor tuning. This includes care and feeding of SOC IT equipment: servers, workstations, printers, relational databases, trouble–ticketing systems, storage area networks (SANs), and tape backup. If the SOC has its own enclave, this will likely include maintenance of its routers, switches, firewalls, and domain controllers, if any. This also may include O&M of monitoring systems, operating systems (OSes), and hardware. Personnel who support this service have "root" privileges on SOC equipment.
Sensor Tuning and Maintenance	Care and feeding of sensor platforms owned and operated by the SOC: IDS, IPS, SIEM, and so forth. This includes updating IDS/IPS and SIEM systems with new signatures, tuning their signature sets to keep event volume at acceptable levels, minimizing false positives, and maintaining up/down health status of sensors and data feeds. SOC members involved in this service must have a keen awareness of the monitoring needs of the SOC so that the SOC may keep pace with a constantly evolving consistency and threat environment. Changes to any in-line prevention devices (HIPS/NIPS) are usually coordinated with the NOC or other areas of IT operations. This capability may involve a significant ad hoc scripting to move data around and to integrate tools and data feeds.

Table 1. SOC Capabilities

Name	Description
Custom Signature Creation	Authoring and implementing original detection content for monitoring systems (IDS signatures, SIEM use cases, etc.) on the basis of current threats, vulnerabilities, protocols, missions, or other specifics to the constituency environment. This capability leverages tools at the SOC's disposal to fill gaps left by commercially or community-provided signatures. The SOC may share its custom signatures with other SOCs.
Tool Engineering and Deployment	Market research, product evaluation, prototyping, engineering, integration, deployment, and upgrades of SOC equipment, principally based on free or open source software (FOSS) or commercial off-the-shelf (COTS) technologies. This service includes budgeting, acquisition, and regular recapitalization of SOC systems. Personnel supporting this service must maintain a keen eye on a changing threat environment, bringing new capabilities to bear in a matter of weeks or months, in accordance with the demands of the mission.
Tool Research and Development	Research and development (R&D) of custom tools where no suitable commercial or open source capability fits an operational need. This activity's scope spans from code development for a known, structured problem to multiyear academic research applied to a more complex challenge.
Audit and Insider Threat	
Audit Data Collection and Distribution	Collection of a number of security-relevant data feeds for correlation and incident analysis purposes. This collection architecture may also be leveraged to support distribution and later retrieval of audit data for on-demand investigative or analysis purposes outside the scope of the SOC mission. This capability encompasses long-term retention of security-relevant data for use by constituents outside the SOC.
Audit Content Creation and Management	Creation and tailoring of SIEM or log maintenance (LM) content (correlation, dashboards, reports, etc.) for purposes of serving constituents' audit review and misuse detection. This service builds off the audit data distribution capability, providing not only a raw data feed but also content built for constituents outside the SOC.
Insider Threat Case Support	Support to insider threat analysis and investigation in two related but distinct areas: 1. Finding tip-offs for potential insider threat cases (e.g., misuse of IT resources, time card fraud, financial fraud, industrial espionage, or theft). The SOC will tip off appropriate investigative bodies (law enforcement, Inspector General [IG], etc.) with a case of interest. 2. On behalf of these investigative bodies, the SOC will provide further monitoring, information collection, and analysis in support of an insider threat case.
Insider Threat Case Investigation	The SOC leveraging its own independent regulatory or legal authority to investigate insider threat, to include focused or prolonged monitoring of specific individuals, without needing support or authorities from an external entity. In practice, few SOCs outside the law enforcement community have such authorities, so they usually act under another organization's direction.

Table 1. SOC Capabilities

Name	Description
Scanning and Assessment	
Network Mapping	Sustained, regular mapping of constituency networks to understand the size, shape, makeup, and perimeter interfaces of the constituency, through automated or manual techniques. These maps often are built in cooperation with—and distributed to—other constituents. For more information, see Section 8.1.2.
Vulnerability Scanning	Interrogation of consistency hosts for vulnerability status, usually focusing on each system's patch level and security compliance, typically through automated, distributed tools. As with network mapping, this allows the SOC to better understand what it must defend. The SOC can provide this data back to members of the constituency—perhaps in report or summary form. This function is performed regularly and is not part of a specific assessment or exercise. For more information, see Section 8.1.3.
Vulnerability Assessment	Full-knowledge, open-security assessment of a constituency site, enclave, or system, sometimes known as "Blue Teaming." SOC members work with system owners and sysadmins to holistically examine the security architecture and vulnerabilities of their systems, through scans, examining system configuration, reviewing system design documentation, and interviews. This activity may leverage network and vulnerability scanning tools, plus more invasive technologies used to interrogate systems for configuration and status. From this examination, team members produce a report of their findings, along with recommended remediation. SOCs leverage vulnerability assessments as an opportunity to expand monitoring coverage and their analysts' knowledge of the constituency.
Penetration Testing	No-knowledge or limited-knowledge assessment of a specific area of the constituency, also known as "Red Teaming." Members of the SOC conduct a simulated attack against a segment of the constituency to assess the target's resiliency to an actual attack. These operations usually are conducted only with the knowledge and authorization of the highest level executives within the consistency and without forewarning system owners. Tools used will actually execute attacks through various means: buffer overflows, Structured Query Language (SQL) injection, and input fuzzing. Red Teams usually will limit their objectives and resources to model that of a specific actor, perhaps simulating an adversary's campaign that might begin with a phishing attack. When the operation is over, the team will produce a report with its findings, in the same manner as a vulnerability assessment. However, because penetration testing activities have a narrow set of goals, they do not cover as many aspects of system configuration and best practices as a vulnerability assessment would. In some cases, SOC personnel will only coordinate Red-Teaming activities, with a designated third party performing most of the actual testing to ensure that testers have no previous knowledge of constituency systems or vulnerabilities. For more information on penetration testing and vulnerability assessment methodology, see the testing sections in [30], [46], [31], [32], [34], and [35].

Table 1. SOC Capabilities

Name	Description
Outreach	
Product Assessment	Testing the security features of point products being acquired by constituency members. Analogous to miniature vulnerability assessments of one or a few hosts, this testing allows in-depth analysis of a particular product's strengths and weaknesses from a security perspective. This may involve "in-house" testing of products rather than remote assessment of production or preproduction systems.
Security Consulting	Providing cybersecurity advice to constituents outside the scope of CND; supporting new system design, business continuity, and disaster recovery planning; cybersecurity policy; secure configuration guides; and other efforts.
Training and Awareness Building	Proactive outreach to constituents supporting general user training, bulletins, and other educational materials that help them understand various cybersecurity issues. The main goals are to help constituents protect themselves from common threats such as phishing/pharming schemes, better secure end systems, raise awareness of the SOC's services, and help constituents correctly report incidents.
Situational Awareness	Regular, repeatable repackaging and redistribution of the SOC's knowledge of constituency assets, networks, threats, incidents, and vulnerabilities to constituents. This capability goes beyond cyber intel distribution, enhancing constituents' understanding of the cybersecurity posture of the constituency and portions thereof, driving effective decision making at all levels. This information can be delivered automatically through a SOC website, Web portal, or email distribution list.
Redistribution of TTPs	Sustained sharing of SOC internal products to other consumers such as partner or subordinate SOCs, in a more formal, polished, or structured format. This can include almost anything the SOC develops on its own (e.g., tools, cyber intel, signatures, incident reports, and other raw observables). The principle of quid pro quo often applies: information flow between SOCs is bidirectional.
Media Relations	Direct communication with the news media. The SOC is responsible for disclosing information without impacting the reputation of the constituency or ongoing response activities.

2.5 Situational Awareness

For a SOC to effectively provide a set of capabilities to constituents, it must understand the environment in which it executes the CND mission, both at a macroscopic and microscopic level. A large portion of a SOC's job, whether intentionally or by accident, is to maintain and provide this understanding of the constituency's defensive posture back out to its constituents. This understanding is referred to as **situational awareness (SA)**.

The most commonly accepted definition of SA in a generic context can be found in Endsley [47]:

Situation awareness is the perception of the elements of the environment within a volume of time and space, the comprehension of their meaning, and the projection of their status in the near future.

The idea of SA grew out of aviation [47, pp. 32-33] in the latter half of the 20th century. Imagine a fighter pilot in his aircraft. Along with his view out of the cockpit window, he has an array of instruments to help him understand where his aircraft is, what its status is, and his surroundings, such as other nearby aircraft. He is constantly making decisions about what to do with his aircraft on the basis of that understanding.

For the fighter pilot to effectively defend himself and his fellow service members against attack, he must be an expert at comprehending a variety of sensory inputs, synthesizing their meaning in the aggregate, and then acting on that understanding. A pilot's priorities are aviate, navigate, and communicate. Similarly, pilots are constantly drilled on three key aspects that comprise their SA: airspeed, altitude, and direction. Their job is complex, and few can do it well.

Ideally speaking, a SOC does essentially the same thing as the pilot, but in the cyber realm. One could argue, however, that the SOC's job is more complex due to the size and complexity of "cyber." While a pilot has one aircraft to control and perhaps no more than a few dozen friendlies or foes around him to keep track of, a SOC may have hundreds of sensors, tens of thousands of assets, and hundreds of potential adversaries. Aviators operate in kinetic space, where instruments normally can be trusted and the results of one's actions are usually obvious. In the cyber realm, analysts must cope with far more ambiguity. The confidence a pilot places in his instruments is high; SOC analysts must always drill down to raw data in an attempt to establish the ground truth of an incident. In fact, sometimes the analyst cannot fully understand exactly what happened, due to incomplete or inconclusive data. Whereas aviation is a topic that has been understood and practiced by over a million people for more than 100 years, CND has been understood and practiced by far fewer for far less time.

> **Ambiguity and uncertainty—the opposites of good cyber SA—exist at every stage of the incident life cycle and must be mitigated through continual improvement in monitoring coverage and analytics.**

The generic definition of SA can be extended to describe **cyber SA** through the Committee on National Security Systems' (CNSS) definition in [42, p. 69], which appears to be derived from Endsley:

Within a volume of time and space, the perception of an enterprise's security posture and its threat environment; the comprehension/meaning of both taken together (risk); and the projection of their status into the near future.

For the SOC, the practice of gaining cyber SA is divided into three components:

1. **Information.** Sensor data, contextual data, cyber intel, news events, vendor product vulnerabilities, threats, and taskings
2. **Analytics.** Interpreting and processing this information
3. **Visualization.** Depicting SA information in visual form.

Section 8.4 and Section 9 discuss the tools that feed information to the analyst as well as SIEM, the cornerstone of analytics. In the cyber SA realm, visualization is still in its infancy. Although there has been extensive work in the area of cyber SA visualization [48] [49] [50], there does not appear to be a universally followed practice for visualizing cybersecurity information.

For the SOC, gaining and using SA follows the observe, orient, decide, and act loop (**OODA Loop**), Figure 2, a self-reinforcing **decision cycle** process originally proposed by John Boyd [51].

The analyst is constantly making observations about the constituency, orienting that information with previous information and experience, making decisions on the basis of that synthesis, taking action, and then repeating the process. Over time spans varying from minutes to years, SOC analysts build their familiarity with their constituency and relevant cyber threats. As that SA is enhanced, they become more effective operators. Because good SA is built up by each analyst over time, staff attrition can be

Figure 2. OODA Loop

a serious impediment to effective SOC operations; therefore, the SOC must take steps to minimize and cope with turnover.

One way to divide up cyber SA is into three related, deeply coupled, and equally important areas: network, mission, and threat. The SOC as a whole must maintain cognizance over all three areas to be effective. These areas of SA are described as follows:

Network

- Number, type, location, and network connectivity of IT assets, including desktops, servers, network devices, mobile devices, and outsourced "cloud" systems

- Network topology, including physical and logical connectivity, boundaries that separate differing zones of trust, and external connections
- Asset, network, and application:
  - Security requirements (confidentiality, availability, integrity)
  - Security architecture (including authentication, access control, and audit)
- State of constituency IT assets
  - What "normal" state looks like across major network segments and hosts
  - Changes in that state, such as changes in configuration, host behavior, ports and protocols, and traffic volume
- The vulnerability of hosts and applications, as seen from different points on the network and network perimeter, and countermeasures that mitigate those vulnerabilities.

Mission

- The lines of business and mission the constituency engages in, including their value, which may be expressed in revenue, expenditures, or lives
- Geographic/physical location where different parts of the mission occur
- The business relationship between the constituency and external parties:
  - The public
  - Businesses
  - Government entities
  - Academic and not-for-profit organizations
- How constituency mission and business processes map to constituency IT assets, applications, enclaves, data, and the dependencies among them
- The role, importance, and public profile of major user groups, such as:
  - System administrators
  - Executives and their administrative staff
  - Those with access to sensitive information (intellectual property, finance)
  - General constituency user population
  - Users external to the constituency
- The meaning of activity on constituency networks and hosts in the context of the mission.

Threat

- Adversaries':
  - Capability, including skill level and resources
  - Intent and motivation
  - Probability of attack

- Level of access (legitimate or otherwise)
- Impact on constituency business/mission and IT
- Likely identity or allegiance
- Actions: in the past, present, and projected into the future
- Assessment and potential attribution of activity to certain adversaries, either internal or external to the constituency.

SA takes on different forms, depending on the level at which cybersecurity decisions will be made. At the lowest tactical level, the network defender sees out to the end asset and enclave, with a direct understanding of hosts and users. Above that, at the operational level, are lines of business and large networks. At the top is the strategic level, where long-term campaigns are waged by the adversary and where entire enterprises exist. As a result, the need to understand the constituency and actors varies widely, depending on what level of SA is appropriate—from the Tier 1 SOC analyst to the CIO and beyond.

The constituency, especially its executives, naturally look to the SOC to answer questions such as, "What's happening on my network?" A SOC can focus on its mission by providing details to constituents and other SOCs, during both normal operations and a critical incident. Without proactively providing enough detail, the SOC either will be marginalized or will constantly field ad hoc data calls. If the SOC provides too much detail, its resources will be overcommitted to answering questions from constituency management and thus unable to adequately spot and analyze intrusion activity.

> **Finding a balance in proactive reporting to constituents and other partner SOCs will help the SOC gain recognition as a valued resource.**

Effective reporting will also spur partner organizations—especially other cybersecurity stakeholders—to provide feedback, reinforcing and growing the SOC's SA.

Having mature cyber SA allows the network defender to answer some crucial questions:

- What is the patch status of the enterprise? Which patches do we really need to care about, and which are less important?
- Is my constituency facing the serious threat of a targeted external attack such as a spear phishing campaign?
- What is a useful, real-time picture of possible intrusions or, at the very least, known malware?
- To which systems should I apply different security controls that will provide the greatest overall help in preventing a given set of attacks?

- What is changing about the threats faced by the constituency? How are their TTPs changing, and what do I have or need to detect and defend against those new threats?
- Who is acting outside their typical lines of behavior, and is this cause for concern?
- What is the relevance of the attacks I'm investigating within the context of the constituency mission?

2.6 Incident Tip–offs

A SOC's number one job is to find and respond to security incidents. Potential incident reports, or "**tippers**," can come from a number of parties, including:

- Constituents with normal, unprivileged system access
- Constituency system and network administrators
- Constituency help desk
- Constituency ISSOs and ISSMs
- Legal counsel or compliance entities
- Peered, subordinate, coordinating, or national SOCs
- Law enforcement or other investigatory entities
- Other organizations somehow involved in the incident.

These tip-offs can be delivered through a variety of methods, typically:

- Email messages
- Phone calls
- Walk-in reports
- Incident reporting form on SOC website
- Cyber tip feeds (from other SOCs).

Incidents reported through these means should usually be regarded as high-value, especially in comparison to unconfirmed IDS alerts. Because they were evaluated in the context of these parties' own missions and systems, they are almost certain to be worthy of attention. That said, these incidents must be carefully evaluated against data the SOC has access to in order to evaluate whether follow-up is necessary. Incidents stemming from external tip-offs only constitute a portion of the SOC's entire caseload. Some, if not most, incidents are detected and run to ground through internal means that are discussed more fully in Section 8:

1. Network intrusion detection system/network intrusion prevention system (NIDS/ NIPS)
2. Host intrusion detection system/host intrusion prevention system (HIDS/HIPS)
3. Network traffic log (NetFlow) collection and analysis

4. Operating system, application, and network device logs such as Web server logs, Web proxy logs, DNS logs, and anti-virus (AV) alerts
5. LM/analysis tools and SIEM that collect, analyze, correlate, and report on these logs
6. Adversary sandboxing and observation techniques ranging from malware "detonation chambers" to honeypots
7. Forensic analysis of malware samples and media images such as hard drives
8. Automated collection and alerting on cyber intel
9. Full-session network traffic capture (PCAP) analysis.

A SOC's focus is detecting with these tools and data, but they can often be expensive to operate and maintain. Typically, educating the constituency is the cheapest way to obtain good incident tip-offs. Smaller, less mature SOCs may often rely heavily on tip-offs from members of the constituency and existing log sources. Larger SOCs, with more resources, can leverage methods to detect incidents on their own, such as a blanket of network and host sensors and SIEM. Even with a well-educated constituency, a SOC still needs significant infrastructure to gather supporting data, perform analysis, and respond.

Ideally, a SOC will detect an incident before significant damage is done. Historically, SOCs have focused their efforts on the ideal case: detecting an incident while the adversary is performing reconnaissance on the target or during the actual attack where a presence is gained on a system through exploiting a vulnerability.

Given the increasing sophistication and obscurity of attacks, SOCs are compelled to consider the entire "**cyber attack life cycle**,"[1] (Figure 3) also referred to as the "cyber kill chain," a concept adapted from traditional kinetic warfare in [52]. Using this approach, the SOC strives to detect and responds to adversaries, not just when they deliver their

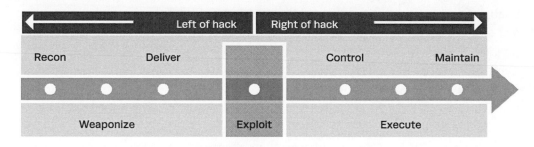

Figure 3. Cyber Attack Life Cycle

---

1 The version of the cyber attack life cycle presented in Figure 3 and Table 2 is derived from the Lockheed Martin version of the cyber kill chain [317].

attack to a target, but throughout the life cycle, from reconnaissance to an extended presence on compromised systems.

Table 2 synopsizes each stage of the cyber attack life cycle.

Using knowledge of the entire cyber attack life cycle, we can take a more holistic approach to sensing and analytics. For instance, sensor instrumentation of constituency networks and hosts should not only provide indications of reconnaissance and exploit activity but should also reveal the presence of remote access tools (RATs) used in the control and execute phases. Furthermore, evidence of an exploit may be elusive, considering that it may not be known (e.g., "0-day" attack) or may not be seen on the network. That said, a typical objective such as sensitive data exfiltration can sometimes be detected through abnormally large amounts of data being transmitted out of the enterprise. Finally, many incidents occur

Table 2. Cyber Attack Life Cycle

Cyber Attack Life Cycle Phase	Description	Example
Recon(naissance)	The adversary identifies and investigates targets.	Web mining against corporate websites and online conference attendee lists.
Weaponize	The set of attack tools are packaged for delivery and execution on the victim's computer/network.	The adversary creates a trojanized Portable Document Format (PDF) file containing his attack tools.
Deliver	The packaged attack tool or tools are delivered to the target(s).	The adversary sends a spear phishing email containing the trojanized PDF file to his target list.
Exploit	The initial attack on the target is executed.	The targeted user opens the malicious PDF file and the malware is executed.
Control	The adversary begins to direct the victim system(s) to take actions.	The adversary installs additional tools on the victim system(s).
Execute	The adversary begins fulfilling his mission requirements.	The adversary begins to obtain desired data, often using the victim system as a launch point to gain additional internal system and network access.
Maintain	Long-term access is achieved.	The adversary has established hidden backdoors on the target network to permit regular reentry.

because trusted parties use legitimate privileges for illegitimate ends (e.g., industrial espionage). As a result, some steps in the attack life cycle may be skipped.

> **The SOC has the best chance of catching the adversary when it equips the constituency with capabilities that cover the entire attack life cycle.**

Finally, the SOC must recognize that evidence of attacks can occur in multiple places—through network traffic or in the host's basic input/output system (BIOS), firmware, hard drive, removable media, system-level software and applications in memory, and less privileged user-level applications in memory. This recognition impacts both what sensors are deployed and the operators' understanding of when and how these sensors can be blinded, disabled, or simply circumvented.

The mantra of "defense-in-depth" [53] applies to incident monitoring and response. From corruption of system firmware to attacks traversing Internet gateways, the SOC must deploy a variety of techniques to cover the entire cyber attack life cycle and constituency.

2.7 Tools and Data Quality

The CND mission succeeds or fails by the SOC analysts' ability to collect and understand the right data at the right time in the right context. Virtually every mature SOC uses a number of technologies that generate, collect, analyze, store, and present tremendous amounts of security-relevant data to SOC members. Figure 4 illustrates the high-level SOC architecture of the tools and technologies a SOC uses.

A SOC may place monitoring tools at many points in the constituency to look for evidence of malicious or anomalous activity across each stage of the cyber attack life cycle. For instance, both key hosts and network choke points provide appealing locations for SOC analysts to look for indications of anomalous or malicious activity. Each tool produces a series of events and raw contextual data that, once examined, may provide evidence of an incident.

In this section, we discuss the fundamentals of how most SOC detection tools operate, and the trade-offs the SOC faces when considering what and how much data to gather in an effort to find intrusions and run them to ground.

2.7.1 From Tip-offs to Ground Truth

The SOC must carefully decide how it instruments constituency IT assets and networks with sensors and log feeds, both for incident tip-offs and supporting context. This concept is nothing new. On the morning of 7 December 1941, a radar station in Oahu, Hawaii,

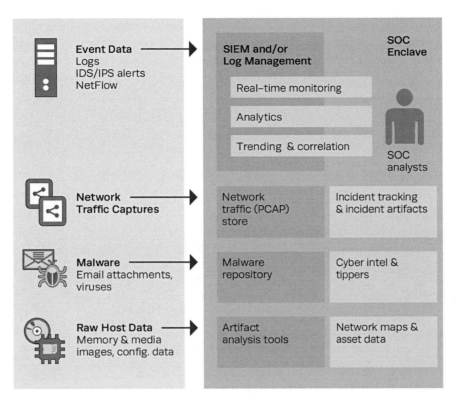

Figure 4. Typical SOC Tool Architecture

operated by the U.S. Army, picked up a huge blip on its instruments. Vigilant radar operators sent news of their finding to another unit on the island; however, they could not
understand what was actually on their radar [54] because they had no other observables
to confirm what they were seeing or provide context. They speculated that the blip was
caused by friendly B-17s flying a mission that same morning. Unfortunately, what was
actually on their radar were Japanese warplanes on their way to bomb Pearl Harbor. Had
the radar operators been able to confirm that the blip on their instruments was the enemy,
they could have taken preemptive measures to prepare for the attack. The same can be
said of many high-severity alerts seen by SOC analysts—without supporting data, the raw
alert is worth little.

> **No matter how severe, a single raw event by itself does not
> provide sufficient evidence that an incident actually occurred.**

Each event must be evaluated within the context of the system(s) it occurred on, the surrounding environment, supported mission, relevant cyber intelligence data, and other sources of data that can confirm or repudiate whether there is any cause for concern. This is why logical and physical proximity between the analysts and the systems they monitor enhances their ability to monitor and respond. Analysts will start with initial indicators (such as a high-priority alert or correlation rule trigger) and gather additional contextual data to evaluate the "ground truth" disposition of the event or series of events they are evaluating. Confidence in this data can be enhanced through techniques such as correlation and automating portions of data filtering and deduplication with SIEM. Often, the SOC cannot be 100 percent certain about what actually happened, due, usually, to incomplete, inconclusive, or ambiguous data. This is a reality that the SOC must cope with, especially when weighing its response options.

While automation will save the SOC from drowning in a sea of raw data, do not forget the following truth:

There is no replacement for the human analyst.

Until a SOC analyst has evaluated the disposition of an alert, the SOC cannot be certain that what occurred is a confirmed incident or is benign. Furthermore, no one analyst can know all the technologies or all the behaviors of the enclaves and hosts he or she is watching; multiple sets of eyes and analytics usually are better than one. There are cases where a set of indicators are correct often enough that certain response actions can be automated—such as with IPSes or SIEM automated response products. However, there is always the chance that a positive indicator will turn out to be incorrect—these are called **false positives** and are discussed in greater detail below.

For now, however, consider tip-offs and contextual data as a spectrum:

Tip-offs help orient the analyst as to what events should be looked at

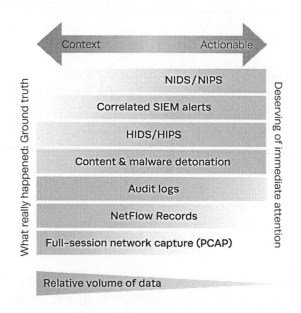

Figure 5. Context to Tip-offs: Full-Spectrum CND Data

first. Context is needed to establish a complete detail of what actually happened (e.g., the undisputable "facts" of what occurred on the host and across the network). When a SOC's monitoring does not cover this spectrum, its analysts will not be able to effectively spot and analyze suspected incidents.

2.7.2 Detection

Several of the tools discussed here—host- and network-based IDSes/IPSes, SIEM, log aggregation tools, and AV software—follow a similar set of characteristics:

1. Knowledge of the environment and the threat is used to formulate a detection policy that defines known good, known bad, or normal behavior.
2. A detection engine consumes a set of cyber observables (e.g., events or stateful properties that can be seen in the cyber operational domain) and compares them against a detection policy. Activity that matches known bad behavior or skews from typical "normal" behavior causes an event or a series of events containing details of the activity to be fired.
3. Events can be sent to the CND analyst in real time or stored for later analysis and trending, or both.
4. Feedback from the events generated will inform further tweaks to the detection policy; this process of authoring and refining the detection policy is known as "tuning."
5. Events can be filtered in several places in a large monitoring architecture, most notably in two places: before the events are displayed to the analyst or before they are stored.

Although IDSes and SIEM operate at different layers of abstraction, they both fit this model. The network IDS observables are network traffic. IDS alerts and logs feed SIEM, which treats these events as cyber observables, as the IDS did.

There are two classical approaches to intrusion detection [2, pp. 86–87]:

1. **Misuse** or **signature-based detection**, where the system has *a priori* knowledge of how to characterize and therefore detect malicious behavior
2. **Anomaly detection**, where the system characterizes what normal or benign behavior looks like and alerts whenever it observes something that falls outside the scope of that behavior.

Both approaches have pros and cons. In practice, finding tools that rely exclusively on one approach or the other is rare; most IDS and AV vendors integrate both techniques into their products. Commercial IDS vendors do this to enhance their products' apparent value; however, the products' effectiveness may vary with the robustness of their detection

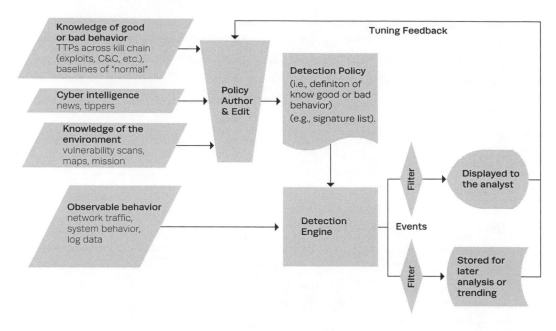

Figure 6. CND Tools in the Abstract

engines. That said, examining most SOCs' tool chests usually reveals a greater reliance on signature-based approaches than on anomaly detection tools. The most important thing to recognize is that signature-based detection requires the defender to know about adversary techniques in advance, which means the tool cannot alert on what has never been seen. On the other hand, training an anomaly detection tool suite often requires something akin to a baseline of "goodness," which can be hard to attain.

2.7.3 The Cost of False Positives

Tools do not always alert when something bad happens, and just because they throw an alarm does not necessarily mean it is time to isolate a host or call the police. Just because an IDS alerted that "the Web server has been hacked" does not mean that the Web server was actually hacked. Each event that detection tools generate falls into one of four categories, depending on whether alert fired and whether something bad actually happened:

1. **True positives.** Something bad happened, and the system caught it.
2. **True negatives.** The activity is benign, and no alert has been generated.
3. **False positives.** The system alerts, but the activity was not actually malicious.
4. **False negatives.** Something bad happened, but the system did not catch it.

The most prominent challenge for any monitoring system—particularly IDSes—is to achieve a high true positive rate. In both academia and commerce, creators and salespeople for monitoring systems strive to prove that their system never misses a successful hack (i.e., it never has false negatives). However, this often comes at the price of a torrent of false positives, as discussed in [55]. Too many false positives compel analysts to spend inordinate amounts of time wading through data or, worse, ignoring a tool entirely because the good signal is lost in the noise.

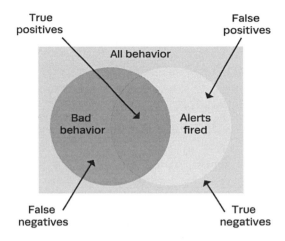

Figure 7. The Four Categories of Activity

On 20 April 2010, a Transocean oil rig off the coast of Louisiana exploded, killing several workers and spilling millions of gallons of oil into the Gulf of Mexico. The disaster and demise of the rig, Deepwater Horizon, cost millions of dollars in direct business and economic damage, not to mention the environmental disaster and cleanup that followed for months and years [56].

A number of factors precipitated this event; we will focus on one—the rig's safety alarm systems. The alarms for one of the safety systems on the rig were disabled because they went off too often. To avoid wakening maintenance personnel in the middle of the night, the alarms were essentially deactivated for a year prior to the explosion [57]. One of the main themes of the Transocean incident was that ongoing maintenance problems prevented the safety systems from operating and alerting correctly. The Transocean incident shows that an analyst becoming numb to even the most serious of alarms can have disastrous consequences. While the Transocean disaster happened for a number of reasons, ignoring safety alarms clearly did not help.

> **False positives often outnumber true positives in any detection system. Leveraging robust detection engines, continual tuning and filtering, and analysis leveraging rich contextual data are critical to ferreting out what is truly of value.**

Labeling a nonmalicious event a "false alarm" presumes that the signature author's intent was to detect some sort of malicious activity; sometimes this is not the intent. SOCs often choose to activate various IDS signatures and accept log feeds because they will leverage the data for contextual or retrospective analysis, not because they serve as tip-offs. A fleet of IDS sensors may generate thousands of "good" tip-offs a day, while millions of audit logs are collected to back them up.

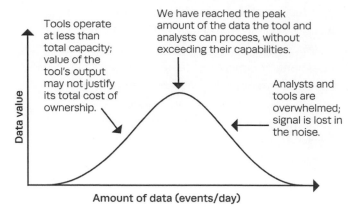

Figure 8. Balancing Data Volume with Value

These low-priority "audit" events are not false positives, considering that their intent was not to indicate "badness."

For example, an IDS may be configured to log all Web requests because the SOC does not have Web proxy logs at its disposal. The resulting events are not false positives; they indicate exactly what was intended—Web requests. Just one IDS sensor or log data source can generate more than 100,000 events each day. Tools such as SIEM facilitate and automate alert triage. As a result, the SOC can actually be more liberal with their signature policies. More frequent events usually are the ones more ripe for correlation, trending, or, perhaps, tuning, whereas the rare events usually are more interesting as tip-offs.

Some IDS and SIEM tools provide complete details of the signature or rules that are used by the system to generate alerts, along with the raw data that generated the alarm. Therefore, one could assert that there is no such thing as a false positive, only an analyst who does not know how to interpret the results. The term "false positive" may be best used in reference to an event generated by a signature whose *intent* was to indicate "badness," but some flaw in its accuracy or precision caused it to alarm under incorrect circumstances. Signature precision is always a challenge, and, although most IDSes integrate the concept of event priority, they do not integrate the concept of confidence. That is, what is the precision and accuracy of the signature relative to the activity of concern?

2.7.4 Data Quantity

Not all data is created equal, and the SOC must prioritize which data to aggregate, process, and store, according to what it expects to get out of that data. The SOC must recognize the limitations of its tools, gauge the detection and analysis use cases it wants to support, and properly size the amount of data it collects. Too little data and the tools are underutilized.

Too much data and two problems exist: (1) the signal is lost in the noise, and (2) the systems cannot handle the data load in terms of ingest or query performance, or both.

Think of it this way. Miners spend tremendous resources digging through dirt, sand, and rock to find precious materials such as gold or diamonds. They must choose the right dirt to sift through to maximize resources spent on finding valuable material in a given day. If they choose to indiscriminately dig through a huge mountain of dirt, it is unlikely that they will uncover the most mineral-rich deposits. Better and bigger tools help them process more dirt, but the principle stands—they must first find the right place to dig.

SIEM can enhance the value of just a few low-volume data feeds, but most COTS SIEMs have a high acquisition and maintenance cost. As a result, it is not cost efficient to deploy a multimillion-dollar SIEM and only bring in 10,000 events a day (considering that Perl scripts and grep might do the job for free). As we discuss in <u>Section 8.4</u>, log aggregation appliances and SIEM are *usually* worthwhile when event rates are measured in millions per day, or more than a handful of data types are available to be gathered.

Figure 9 provides another way to look at the data ingestion challenges.

FOSS tools such as Perl scripts, grep, and the like provide a number of interesting ways to slice and dice log data, especially when the operator must handle thousands or perhaps a few million a day. However, moving along the chart toward the right suggests that larger, more diverse data sets will gain more value by leveraging log collection and SIEM's framework and analytics. Many SOCs deal with event rates measured in the tens or hundreds of millions per day, in which case they will typically choose an LM appliance or a SIEM tool, or a combination thereof. The SOC also must weigh trade-offs between custom tools that provide more flexibility and commercial tools that have vendor support, all while factoring in the total cost of ownership (TCO).

The "collect everything" strategy is not a good one when we have limited analysts, hard drive space, and central processing unit (CPU) cycles. For instance, many organizations are subject to an audit policy along the lines of "record access or modification to key information." Defining "key" is universally problematic and usually ends up encompassing almost any system or database of consequence. The policy then is often interpreted as "generate and collect audit records for every file read/write

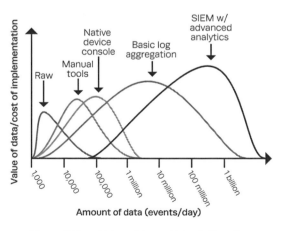

Figure 9. Data Volume Versus Value in Practice

on every system in the enterprise." These audit records will dwarf most other data feeds, obligating system resources and crushing every audit collection system in its path. The moral is this:

> No matter how good the tool or analyst, overzealous efforts to generate and aggregate huge amounts data into one place diminish the value of good data because it is lost in the noise of worthless data.

The old saying "looking for a needle in a haystack" is commonly used to refer to the bulk of CND monitoring, especially when analysts lack either effective data reduction and correlation tools or context for the data they are examining. Some SOCs collect as much useful, security-relevant data as they think will be relevant ahead of time, and mine it later for actionable indicators. This strategy works for SOCs that have storage, processing power, and analyst cycles to spare, which, of course, is not always the case.

Operating a fleet of monitoring capabilities requires the SOC to be constantly vigilant because the threat landscape and the makeup of the constituency are constantly changing. Data feeds go down, sensors become unplugged, new signatures come out, and so forth. Many SOCs lose sight of these facts, which is perhaps the main reason that IDS and SIEM deployments often are seen as lacking value. One of the most important aspects of monitoring the constituency is:

> Monitoring systems such as IDS and SIEM are not "fire and forget"—they require regular care and feeding.

These themes are discussed later in the book, especially where we discuss instrumenting the enterprise and where we consider how to detect the APT.

2.8 Agility

If the SOC has a worst enemy, some might say that it is the APT. The adversary acts, reacts, and changes strategies quickly—its version of the OODA Loop decision cycle is usually tightly wound. How fast do large organizations such as government agencies or Fortune 500 companies act in cyberspace? Usually not in seconds. In many cases, the SOC must expend all of its effort just to maintain the same pace of advancement as the adversary; this is further complicated by a constantly changing constituency enterprise. The SOC must expend even more effort to actually advance its capability. This fact is highlighted in the cover of this book.

Consider the issues we must wrangle with: design by committee, change control, ambiguous or conflicting lanes of responsibility, slow network performance, decentralized control, and laborious policies and procedures. Of all the things that undermine the SOC's ability to spot and repel attacks, most often it is the lack of speed and freedom with which the SOC may act—a lack of *agility*.

> **The key to effective CND is having the people, process, and technology that enable the SOC to maintain parity with the adversary.**

The best SOCs stand out in a number of ways, but a high operations (ops) tempo is one of the most prominent. Let's compare the adversary's decision cycle with the ops tempo of an effective SOC with a constituency of around 20,000 hosts and users. In Table 3, the Attacker column outlines what today's APT actors can do in given time ranges—from years to seconds. The Defender column outlines what the defender must do to maintain parity with the attacker, within the same sorts of time spans. The Attacker column is *descriptive*, while the Defender column is *prescriptive*—not how it is but *how it should be*.

Table 3. Ops Tempo of the Attacker Versus the Defender

	Attacker	**Defender**
Years	Evolve new classes of attacks.	Make new defense techniques operational. Gather and execute funding, contracting, and technology deployment for a new SOC.
Months	Develop teams of participant script kiddies or thousands of "bots" able to target multiple large enterprises. Execute an entire intrusion campaign against a large, complex target such as a Fortune 500 company or a government agency.	Conduct a technical bake-off among multiple competing products. Recap network or host sensor fleets. Fully instrument a data center with monitoring coverage. Hire IT professionals and train them as Tier 1 CND analysts. Draft, socialize, finalize, and sign enabling policy documents and authorities.
Weeks	Perform detailed reconnaissance on a targeted organization. Develop local or remote exploits against an application, from scratch. Exfiltrate terabytes of sensitive data.	Develop, deploy, and make operational complex custom detection and analytics tools such as Perl scripts and SIEM use cases. Revise, review, and baseline an internal SOC standard operating procedure (SOP).

Table 3. Ops Tempo of the Attacker Versus the Defender

	Attacker	Defender
Days	Once inside, thoroughly enumerate and take over an entire enterprise. Weaponize a patch release into an attack vector.	Do a monthly/quarterly scrub of all signatures deployed to an IDS fleet or content deployed to SIEM. Test and push a major patch to an enterprise. Analyze and document the contents of a hard drive image from a system involved in a serious incident. Assess the actions and potential motives and intentions of an adversary operating on constituency networks.
Hours	Once inside an enterprise, escalate from user to administrative privileges.	Develop or download and deploy IDS signatures to a fleet of sensors. Identify, analyze, and develop a response plan to an intrusion involving multiple systems or accounts. Provide cursory analysis of the payload for a new strain of malware. Identify and recover from a downed sensor or data feed. Gather stakeholders and brief them on details of a major incident in progress.
Minutes	Phish a large enterprise user population. Establish a covert sustained presence on a host. Exfiltrate targeted sensitive data from key assets.	Query each month's log data for any system in an enterprise and gather results. Retrieve a week's worth of indexed PCAP from online storage for a given set of IP addresses. Recognize an event of concern and tag it as benign or fill out a case and escalate it to Tier 2. Isolate an infected host. Identify and contact a sysadmin at a site whose system was involved in a potential incident. Automatically extract an email attachment from the network, execute it in a detonation chamber, and analyze it for signs of malicious activity.

Table 3. Ops Tempo of the Attacker Versus the Defender

	Attacker	Defender
Seconds	Compromise a host through drive-by downloads or remote exploitation. Hop from one stepping-stone IP or domain to another, circumventing IP block lists. Morph specific attack code, thereby circumventing signature-based IDS. Exfiltrate a handful of highly sensitive documents from a website or network share.	Automatically prevent an attack at the network or host through a protective tool such as HIPS. Generate an audit entry and send it to a SIEM console. Trigger an IDS alert and send both the alert and the associated packet to the SIEM console.

Consider organizations that you are familiar with. Can they move with this kind of speed? Even if they can, gaps still exist between how fast the attacker can move and how quickly the defender decides the best response approach (e.g., the difference between how fast an attacker can hop IPs or domains and how fast the defender can block it). Keep these time spans in mind in later sections of this book as to how people, process, and technology are marshaled to enable effective network defense.

In the following ten sections, we use the foundational matter we have just discussed to cover the ten strategies of world-class SOCs in detail.

Chapter 3
STRATEGY 1: Consolidate CND Under One Organization

The first strategy is the most obvious but the least often followed: consolidate functions of CND under one organization whose sole mission is executing the CND mission. As discussed in Section 2.8, SOCs must be able to respond in a time scale relevant to the actions of the adversary. As a result, elements of CND must be tightly coupled. Bringing the CND mission into a single organization makes possible the following goals:

- Operations are synchronized among the elements of CND.
- Detection and response are executed efficiently, without sacrificing accuracy, effectiveness, or relevancy.
- Resources spent on CND can be maximized.
- Cyber SA and incident data is fed back into CND operations and tools in a closed loop.
- Consolidated, deconflicted SA is provided to the director of incident response and his/her management chain.

As a result, we recognize five indivisible, atomic elements of CND that should be under one command structure:

1. Real-time monitoring and triage (Tier 1)
2. Incident analysis, coordination, and response (Tier 2 and above)
3. Cyber intel collection and analysis
4. Sensor tuning and management and SOC infrastructure O&M
5. SOC tool engineering and deployment.

As we describe in <u>Section 4.2</u>, there are many different ways to bring these under one roof; that said, here is one typical organizational structure with functional responsibilities: Unfortunately, contact with several dozen government and commercial SOCs reveals

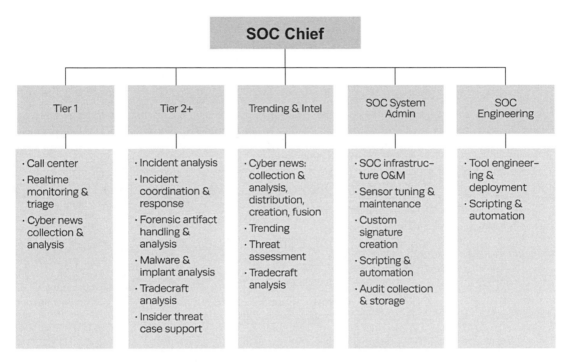

Figure 10. All Functions of CND in the SOC

that CND is frequently broken into multiple independent organizations. For instance, we might see Tier 1 IDS monitoring in a NOC, but Tier 2 incident response located in the office of the CIO.

When we divide up these core functions, we typically see one or more of the following: (1) depressed ops tempo, (2) broken or ineffective internal processes, (3) ineffective communication, (4) slow improvement in mission capability, and (5) animosity/distrust among different parties supporting CND. The result, of course, is a subtle yet profoundly negative impact on the defensive posture of the constituency. Consider the consequences of removing each of the core functions of the SOC:

- Real-time monitoring and triage or incident analysis, coordination, and response:
  - The incident escalation and follow-up process is disjointed and fragmented.
  - Incidents are slow to be followed up, and Tier 1 receives little feedback.

- Quality control of what comes off the ops floor is hard to correct.
- Career progression, and thus retention, of analysts is stunted.
- In some scenarios, CND monitoring architecture and tool set are badly fractured, usually due to subtle differences in user requirements and decentralized resourcing.
- Cyber intel collection and analysts:
  - Nothing drives focus or improvement to SOC monitoring.
  - The SOC does not keep pace with the current threat environment.
  - The SOC is not viewed as resource for cyber SA by constituents.
  - Monitoring tools poorly leverage TTPs and indicators from available cyber intel sources and, thus, do not maintain parity with current threats and vulnerabilities.
- Sensor tuning and management:
  - SOC systems fall into disrepair.
  - Downtime of SOC systems is prolonged because responsible personnel are not accountable to SOC management.
  - Sensors lack current signatures.
  - Monitoring and incident results do not drive improvements or customizations to signature packages or SIEM content.
  - The SOC cannot maintain effective separation and protection from the rest of the constituency because its sysadmins and/or tools are thrown into the general pool of IT support.
  - Systems go down, and SOC personnel's hands are tied when applying tactical corrective actions or resolving larger system design issues.
- SOC tool engineering and deployment:
  - Intense distrust develops between SOC and engineering because of overlapping technical knowledge, but divided priorities and vision.
  - Engineering does not match operational priorities of SOC, both in terms of timely delivery of new capabilities and the needs of the operator.
  - The artificial division between ops and engineering causes confusion over responsibilities; the separate process restricts the SOC from leveraging tactical solutions.
  - Because the engineers are not embedded in ops, they do not fully understand the operators' needs, regardless of how robust and detailed the requirements specification.
  - Many SOC capabilities are best developed in a spiral fashion, providing immediate benefit to ops, and can change as the mission demands it; engineering life cycles where projects are "thrown over the fence" do not support this.

- Ambiguity over funding sources for SOC capabilities can make coordinating program funding, capital expenditures, and maintenance budgeting problematic.
- Fractured CND program budgeting can introduce an imbalance in money for tools, staff, and training.

Based on our discussion from Section 2.8, one can see how splitting up these functions of the SOC can have overwhelmingly negative consequences. SOCs that are fractured in this manner *might* achieve some success if three compensating measures are enacted:

1. The removed capability resides in a single organization; that is, if SOC infrastructure O&M is pulled out, it is assigned to another group whose sole responsibility is that job.
2. The managers of respective organizations (such as SOC ops and engineering) have an excellent working relationship, mutual respect, and constant communication with one another.
3. The organization that operates the capability pulled away from the SOC is still accountable to it through policy, procedures, and, perhaps, a contractual relationship.

Many SOC implementations separate tool engineering and some aspects of system administration from the SOC. This is especially problematic for reasons not recognized by many outside the CND practice. As we will discuss in Section 7.1, two prerequisites for being an effective analyst are a strong background in programming and in network/system administration. As a result, a SOC will naturally have the expertise necessary for SOC tool engineering and deployment in house. If they are not allowed to perform this function for process reasons, it breeds intense frustration and animosity between engineering and ops.

The CND ops tempo demands solution delivery in days or weeks, not months or years. One reason often cited for pulling engineering out of the SOC is to enforce CM. This is a fallacious argument, considering that robust CM processes can (and should) be implemented entirely within the SOC, along with code review and system hardening. Another argument made is that ops and engineering fall under separate lines of business. Because CND demands a unique mind-set and skill set, there is an argument to be made for excepting the SOC from this organizational construct. Also foreshadowing Section 7.1, CND personnel are not interchangeable with staff from other areas of IT, further bolstering this argument.

> **Do not break apart the five atomic SOC functions into disparate organizations; this will almost always work to the detriment of the CND mission.**

In fact, bringing these functions into one organization (the SOC) usually isn't enough—they also should be physically collocated. For instance, if engineering is located in Omaha and ops is located in Atlanta, collaboration and mutual support will likely suffer. Personnel supporting these capabilities should collaborate on a daily or weekly basis, a topic we return to in Section 4.2.

Chapter 4
STRATEGY 2: Achieve Balance Between Size and Agility

SOCs serve constituencies of almost every size, business function, and geographic distribution. The SOC's structure must correspond to that of its constituency, balancing three needs:

1. The need to have a cohesive team of CND specialists
2. The need to maintain logical, physical, or organizational proximity to the assets being monitored
3. The budgetary and authority limitations inherent in the constituency served.

In our second strategy, we seek to strike a balance among these competing needs. As a result we have three closely linked choices to make:

1. What SOC organizational model is the right fit
2. How to place SOC functions into sections with line managers and command structure
3. Where to physically locate members of the SOC, and how to coordinate their activities.

In order to realize this second SOC strategy, we will cover each of these choices in turn. In this section, we also lay out a number of concepts that we build upon in later sections.

4.1 Picking an Organizational Model

4.1.1 Drivers

Key drivers for determining which organizational model is best for the enterprise include:
- Size of constituency, in terms of users, IP addresses, and/or devices
- Frequency of incidents
- Constituency concerns for timeliness and accuracy of incident response.

Size of constituency is both a driver and a challenge. We need to build up a group of well-resourced CND professionals, but they need to maintain visibility out to the edge, operate within the decision cycle of the adversary, and maintain resourcing levels proportional to the constituency's IT budget. For instance, in larger constituencies, the desire to build a team that covers the whole enterprise may be overshadowed by the team's resulting inability to maintain mission relevancy and agility.

> **The further an analyst is separated from monitored assets—logically or physically—the less he/she is able to maintain context and sense of what is normal and abnormal behavior on those hosts and networks, and able, therefore, to respond in a relevant or timely manner.**

This fact is absolutely key to understanding how best to structure security operations in large enterprises. Without strong SA and operational agility, even world-class analysts and tools will be of little value. Luckily, we can use the organizational models discussed in Section 2.3.2 to help us resolve these competing needs.

A team of analysts can maintain familiarity with only so many assets and enclaves. Our goal here is to structure our analysis resources in a way that they can do that while still operating as one team, working toward a common set of objectives with a synchronized ops tempo. Most CND practitioners are used to working in a paradigm where they have direct access to raw data and can directly impact the assets they are monitoring. The critical issue is: how do we do this with larger and larger constituencies in a relevant, meaningful,

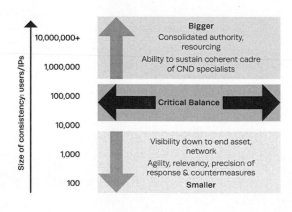

Figure 11. Rightsizing the Constituency

and productive fashion? As of the writing of this book, the answer to this question is not widely agreed upon.

4.1.2 Typical Scenarios

Using what we have discussed so far, we will create five SOC "templates" that we will use throughout the rest of the book, for illustrative and demonstrative purposes. They are shown in Table 4.

Table 4. SOC Templates

1. Virtual SOC	
Organizational Model	Internal Distributed SOC.
Constituency Size	1,000 users/IPs.
Visibility	None/poor. Limited, ad hoc postmortem log review.
Authority	No reactive, no proactive: authorities to prevent or respond to incidents are often vested in the SOC's parent organization.
Examples	SOCs serving small to medium-sized businesses, colleges, and local governments.
Remarks	Comprised of a decentralized pool of resources, this SOC most likely operates out of the office of the CIO, office of the CISO, or in the NOC (if one exists). Incidents do not occur often enough to necessitate constant monitoring.
2. Small SOC	
Organizational Model	Internal Centralized SOC.
Constituency Size	10,000 users/IPs.
Visibility	Limited to good. Instrumentation across major perimeter points, some hosts, and enclaves.
Authority	Shared reactive, shared proactive: the SOC is a voting member in decisions that drive preventative or responsive actions.
Examples	SOCs serving medium-sized businesses, educational institutions (such as a university), or government agencies.
Remarks	Resources for security operations are consolidated under one roof. However, the size of the SOC's budget is limited due to the size of the constituency. If part of a larger organization such as a commercial conglomerate or a large government department, the Small SOC may report to a Tiered or National SOC.

Table 4. SOC Templates

3. Large SOC	
Organizational Model	Internal Centralized SOC, with elements of Distributed SOC.
Constituency Size	50,000 users/IPs.
Visibility	Comprehensive. Instrumentation across most hosts and enclaves.
Authority	Full reactive, shared proactive: the SOC can enact tactical responsive actions on its own, and carries weight in recommending preventative measures.
Examples	SOCs serving Fortune 500 [58] and Global 2000 [59] companies and large government agencies.
Remarks	This SOC is large enough to support advanced services performed from a central location, but it is small enough to perform direct monitoring and response. In more heterogeneous or geographically dispersed constituencies, the Large Centralized SOC may leverage a "hybrid" arrangement with some staff at remote sites for some monitoring and response functions.
4. Tiered SOC	
Organizational Model	Internal Combined Distributed and Centralized, blended with Coordinating SOC.
Constituency Size	500,000 users/IPs.
Visibility	Varies. Some direct data feeds from end assets and enclaves; most data goes to subordinate SOCs.
Authority	Full reactive, shared proactive: the SOC can enact tactical responsive actions on its own, including those that may impact subordinate SOCs, and carries weight in recommending preventative measures.
Examples	SOCs serving multinational conglomerates and large, multidisciplined government departments.
Remarks	Due to the size of the constituency, this SOC has multiple distinct SOCs. There is a central coordinating SOC with its own directly monitored assets and enclaves, most likely located at or near the constituency headquarters and the constituency Internet gateway. There are also multiple subordinate SOCs that reside within given business units or geographic regions, whose operations are synchronized by the central SOC.

Table 4. SOC Templates

5. National SOC	
Organizational Model	Coordinating SOC.
Constituency Size	50,000,000 users/IPs, represented by constituent SOCs.
Visibility	Limited but widespread. No or limited access to raw data *by design*; depends entirely on incident reporting from constituent SOCs; does not directly monitor constituency.
Authority	No reactive, no proactive: despite its powerful name, it is atypical that a national-level SOC can exert substantial authority over its constituents; usually it acts in an advisory role.
Examples	SOCs serving entire national governments or nations.
Remarks	This is a classic national-level SOC that supports dozens to thousands of SOCs within its borders, across governmental, corporate, and educational institutions. Either it does not perform direct monitoring, or, if it does, it provides tippers to its constituent SOCs for follow-up. We will sometimes refer to these organizations as "mega-SOCs." Constituent SOCs operating within the mega-SOC's constituency operate mostly autonomously, which sets this model apart from the Tiered model.

With the first three templates, the SOC is able to maintain direct contact with the constituency, due to its modest size. The last two templates must use sophisticated approaches to support SA to the edge while coordinating CND operations in constituencies of progressively larger sizes. In Section 4.3, we will examine strategies for achieving these goals.

4.2 Structuring the SOC

In this section, we take two of the SOC templates from Section 4.1 with potential capabilities from Section 2.4 to construct a few typical SOC organizational charts. This should give the readers some ideas on how to structure their own SOC to better support smooth operations, without getting into every permutation of what a SOC might look like.

Some strategies we use in structuring the SOC are as follows:

- Put analysts in roles where they function best, but have room to grow both their own capabilities and the SOC mission.
- Maintain separation of duties and eliminate single points of failure to the maximum extent possible.
- Synchronize elements of CND operations so all elements are working in concert toward the same goal, especially during a critical incident.

- Balance energy spent on "managing" with resources devoting to "doing."
- Support the SOC's intended range of capabilities.

4.2.1 Small SOC

Smaller SOCs, in the range of five to 20 people, often find a relatively simple approach to arranging their staff. This is because with few people, there is comparatively less diversification of roles and there are few positions that don't involve full-time analyst work. A classic Small SOC will include two or three sections:

1. Tier 1. Includes analysts who perform routine duties such as watching IDS or SIEM consoles, collecting cyber news, and fielding phone calls from constituents
2. Tier 2. Performs all in-depth analysis on incidents passed to it by Tier 1 such as log and PCAP analysis, and coordinates response to incidents with constituents
3. System administration. Maintains SOC systems and sensors, which may include engineering and deployment of new capabilities.

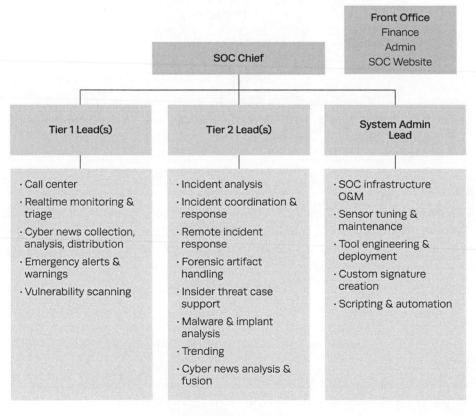

Figure 12. Small SOC

A SOC with this structure can serve a modestly sized constituency while separating "frontline" analysis from in-depth analysis and response. It is shown in Figure 12.

Some SOCs that enforce a tiered analysis structure do not necessarily split the tiers into separate sections. As a variation on the structure shown in Figure 12, we could combine all Tier 1 and Tier 2 duties into one large section with a single operations lead. In this arrangement, we would have two teams—operations and system administration. The other benefit of this setup is that the operations lead can also function as a deputy SOC lead.

Most SOCs of this size have a hard time pulling Tier 2 analysts away from the daily grind of processing incidents. In addition, there are many incidents that stem from activity that does not fall into the structured use cases handed to Tier 1. If this is not corrected, the SOC will likely suffer from stagnation and increased turnover. Foreshadowing Section 11, some SOCs have found it valuable to establish a separate "advanced capabilities" section, shown in Figure 13. This section's roles may vary, but usually incorporate functions such as "Tier 3+" incident analysis, process improvement, and advanced threat detection/

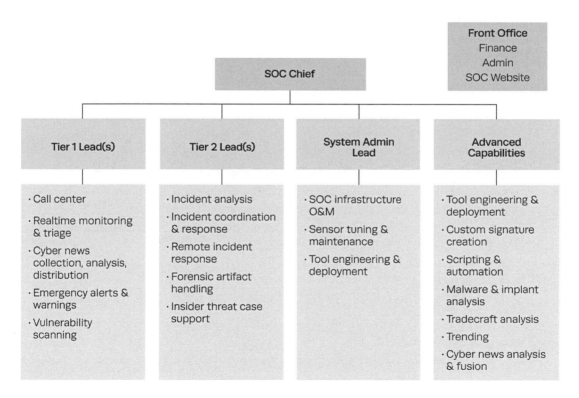

Figure 13. Alternative Arrangement for a Small SOC

response. The advanced capabilities team can be composed of staff (pulled from Tier 2) who demonstrate initiative and out-of-the-box thinking and can be rotated in and out of duties that place them in the "daily grind."

4.2.2 Large SOC

A large constituency can support a SOC with an advanced set of capabilities and full-fledged division of roles and responsibilities. Following our discussion of the Small SOC from the previous section, we have added the following features:

- Tier 1 focuses on fielding phone calls and catching real-time alerts and warnings in the SIEM or other sensor console(s), as a Small SOC does.
- Tier 2 focuses on running incidents to ground, regardless of whether it takes hours or months, as a Small SOC does.
- We have a new section that is responsible for ingesting trending cyber intel and analyzing network activity and adversary TTPs over months and years.
  - This job is often the most ambiguous because analysts are asked to look for open-ended, unstructured threats not currently on the radar.
  - This section is best staffed by self-starters and out-of-the-box thinkers (e.g., the "rock stars" mentioned in Section 7.1.1).
- We have added a host of new capabilities and created a new section that performs both routine network and vulnerability scanning and Blue/Red Teaming for constituency networks and systems.
- O&M and engineering of SOC systems have been divided into distinct groups under one shop, "Systems Life Cycle."
- Within the system administration shop, we will likely have one or two people devoted to each of the most important sensor packages and SIEM.
- The SOC is large enough that it usually has a dedicated deputy position, which may or may not be in addition to the role of ops lead.
- Some SOC chiefs will find it useful to designate middle-level managers in each functional area—analysis and response, scanning/assessment, and system life cycle.
- The "front office" may be added to take away administrative, budgeting, or CM burdens from SOC leads.
- Although not pictured, if the SOC chooses to integrate maintenance of perimeter protection devices such as firewalls, this can be integrated under the systems life cycle lead as a third team.

It may be possible to achieve the same separation of duties for these functional areas in the Small SOC model, but in doing so some staff may be "pigeonholed" into one role, increasing the adverse impact of staff turnover. On the other hand, in the Large SOC, almost every core function is carried out by two or more people.

A potential organizational model for a Large SOC is depicted in Figure 14.

When a SOC gets this large, it is important to ensure there is effective cross-training and cross-pollination. Engineering must stay cognizant of the ops group's main challenges and "pain points" and how to quickly leverage 90 percent solutions. As a result, we can rotate personnel into engineering and development positions, as we did with the advanced capabilities team in the Small SOC model. Moreover, even though we may have multiple layers of management, operators in one section should not hesitate to work directly with any other part of the SOC.

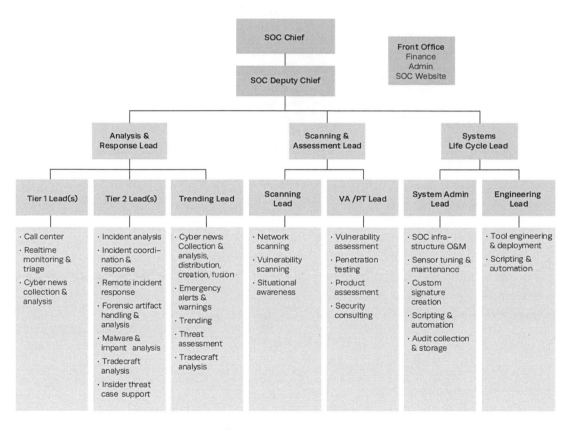

Figure 14. Example: Large SOC

4.3 Synchronizing CND Across Sites and Organizations

The SOC can find its physical location a great help or hindrance, depending on a number of factors. In Section 4.1.1, we talked about the balancing act between SOC size and the need to maintain closeness to the end asset and mission. This may compel us to place SOC personnel at multiple sites, as in distributed or tiered organizational models. In this section, we address the following intertwined issues:

- Where the SOC should be physically placed
- How to arrange SOC resources distributed among several locations
- How to split out duties between a central coordinating SOC and subordinate SOCs, all within one large organization
- Suggested roles and responsibilities for national coordinating SOCs.

In Section 3, we mentioned how the five atomic functions of the SOC should never be broken apart into separate organizations. While we should centralize CND organizationally, we may elect to distribute it physically. So we will offer an important corollary to the previous point:

> **Close physical proximity is instrumental in maintaining a synchronized ops tempo and priorities among parts of the SOC.**

While we have plenty of affordable real-time telecommunications capabilities—Voice Over Internet Protocol (VoIP), video teleconferencing, real-time chat, and desktop webcams—it is rare in an operations shop that we can use them to completely replace physical presence. When different sections of the SOC are moved apart—even to different rooms in the same building—collaboration can suffer. For this reason, we often mix elements of centralized and distributed organizational models, such as leveraging "forward deployed" analysts to reinforce Tiers 1 and 2 at the main SOC ops floor.

4.3.1 Goals and Drivers

Let's highlight some of the needs that we want to meet in making decisions on where to physically place SOC resources. (See Table 5.)

We will use these drivers in the following sections to examine some strategies for the SOC templates from Section 4.1.2.

Table 5. Considerations for SOC Placement

Goal	Discussion
Provide the SOC with a physical space that meets the SOC's mission needs	With the exception of virtual SOC, a physical operations floor will be needed. This usually entails an ops floor, back offices, and server room, all with secure access. This also means having ample bandwidth to constituency wide area networks (WANs), campuses, and data centers. Existing constituency office facilities and data centers may meet these needs better than other options.
Synchronize operations among the sections of the SOC	The atomic functions of CND must be brought into one organization, the SOC. It is also highly useful to bring them under one roof, supporting regular, healthy, and usually informal collaboration.
Maintain clear lines of separation between SOC functions and IT and cybersecurity functions	The SOC sits at the center of a political vortex—plenty of other people in the constituency believe that some element of CND is in their swim lane—fueling conflict with the SOC. Physical presence near these other entities gives the SOC an advantage by supporting close, ongoing contact that can help keep the SOC's interests and impact visible to other stakeholders.
Keep close contact with constituency leadership and other groups from Appendix B	The SOC must leverage support from parent organizations and coordinate with various groups in response to incidents, especially major ones. While this can be done virtually, it's best if they can be brought together physically.
Provide analysts better constituency mission and operations context, speeding analysis and response efforts	Being at a site where IT assets and users reside automatically gives the SOC many advantages—the ability to interact with constituents, perform touch labor on sensors, and execute on–site incident response.
Ensure the SOC's focus on the constituency is not biased to any one organizational or geographic region	Analysts will tend to automatically tune into what is going on at the site where they are located. This is both a blessing and a curse. If analysts are biased toward their site, what are they missing at the others? In more extreme cases such as large, heterogeneous enterprises, remote sites will perform their own security monitoring and incident response functions without coordinating with the SOC—in large part because they feel the SOC is out of touch with their mission.
Better position the SOC for staff hours that mirror when the constituency is open for business	The SOC's business hours should encompass those of the constituency. It helps to place the SOC in the same time zone as a plurality of the constituency. If the constituency's users and IT assets are spread all over the world, the SOC may have more options to maintain 24x7 operations while keeping analysts employed only during the daytime.
Ensure continuity of operations of the CND mission through geographic diversity of SOC assets	The SOC should ideally be considered integral to the constituency's mission. This may compel SOC management and constituency executives to create one or more additional redundant or "load balanced" operations floors, giving the SOC some geographic diversity and resiliency.

4.3.2 Where to Place the Main SOC

In theory, the SOC could operate from any location that has ample rack space, office space, and connectivity to the rest of the constituency. If the constituency has consolidated its IT into one or a few data centers, the SOC could operate there, providing on-site response for a large proportion of incidents. Doing so would also allow the SOC to orient toward mission systems, enabling them to focus more on what's going on with the computing environment and less on routine politics. In practice, this isn't always the best strategy.

Practically speaking, most SOCs are members of their own constituency. Furthermore, they rarely have absolute authority in incident prevention or response. In this regard, their most important contact(s) are those from whom the SOC derives power, such as the CIO. The SOC must maintain continual contact with constituency seniors in order to stay relevant; this is a distinct characteristic in comparison to IT or network operations.

> **The best place for the SOC is at or very near the constituency headquarters.**

SOC representatives will likely need to meet with key constituency technical points of contact (POCs) (CISO, sysadmins, security personnel, etc.) on a regular basis, and also with constituency seniors (chief technology officer [CTO], chief operating officer [COO], CIO, CEO, etc.) from time to time. There are constant changes to policy, monitoring architecture, threat, and incidents, all of which require regular coordination. If there is insufficient power, space, and cooling for SOC servers or no suitable place for a SOC operations floor in the headquarters building, it may be better for the SOC to find a suitable space at a nearby office building, preferably one already owned or leased by the constituency.

4.3.3 Small and Large Centralized SOCs

If we pursue a centralized SOC model, we must have a way to support a presence at remote sites for purposes of incident response, equipment touch labor, and general visibility. This is crucial when the constituency headquarters is far from major elements of constituency operations. Here are some compensating strategies for a centralized SOC model with a geographically dispersed constituency:

- Have at least two designated POCs or "trusted agents" (TAs) at each major location where the constituency operates. These trusted agents:
  - Are usually sysadmins or security officers (ISSOs)
  - Watch over security-relevant issues at the site, such as new system installs and changes to network architecture

- Hold the keys to SOC racks or rack cages and are the only people who are allowed to physically touch SOC systems
- Are the default contacts for on-site incident response
- Are customers of the SOC's audit collection/distribution capability, if one exists
- Serve as champions for SOC interests at the site.

- Make contact with site TAs at least quarterly to ensure they're still in the same position and that their contact information is still current. Having multiple TAs at a site will help ensure that if one person leaves, the alternate TA can find a suitable replacement.
- Have SOC representatives participate in IT CM/engineering boards for IT assets that operate at remote sites
- Send SOC representatives to quarterly or annual collaboration forums run by IT people at sites where they discuss major initiatives in site IT
- Keep up-to-date rack diagrams for all SOC equipment, both both local or remote
- Have access to updated network diagrams of site networks and enclaves.

As we can see here, the line between centralized and distributed SOC models may appear to blur when we talk about how to keep tabs on remote sites. The main distinction here is that the site TAs don't work for the SOC as their main job. Therefore, the SOC cannot heavily task them outside the scope of incident response and sensor touch labor. In hybrid and distributed models, this is not the case, as we describe in the next section.

4.3.4 Incorporating Remote Analysts

Taking our model of TAs one step further, we can actually deploy SOC personnel to remote sites, thereby augmenting resources at the central SOC operations floor. While these individuals report to the SOC, the SOC's main analysis systems are still near the operations floor, and most incident calls are routed to the ops floor. However, we now have people who perform all the roles of the TA, above, make CND part of their day job, and are accountable to SOC leadership.

Keeping members of the SOC working in concert while spread across multiple sites will certainly be a challenge. Here are some tips on how to keep the whole SOC in sync:

- Ensure that analysts at remote sites go through the same personnel vetting and indoctrination process as all other SOC analysts.
- The SOC CONOPS and escalation SOPs need to support site escalation and response coordination with SOC operations leads. We don't want anyone at the site taking response actions without the knowledge of SOC leadership.
- Consider hiring analysts at remote sites who previously held IT security–related jobs at that site, thereby leveraging their familiarity with local operations and IT "culture."

- Folks at remote sites may get bored and feel disconnected from the main SOC. Some ways to mitigate this are:
  - Bring them back to the SOC for one to three weeks every year, as budget allows, for team cross-pollination and refresher training.
  - Consider having a "virtual ops floor" where all floor analysts and site analysts join an open chat room, video session, or VoIP session while on duty.
  - Call extra attention to successes by site analysts to the rest of the SOC team.
  - Schedule regular visits and telecons by SOC leadership to analysts at remote sites, giving them "face time" and keeping leadership abreast of site activity.
- For sites that host more than a few analysts, consider a "mini ops floor"—perhaps a small set of cubes where site SOC personnel can interact.
- Consider keeping site analysts on the job during their site's business hours.
- Ensure all SOC data feeds and sensors are integrated into one unified architecture. While the site should have its own specific source of log data and monitoring systems, this should be part of one unified, coherent architecture, with analytics tailored to that site or region.
- Some site analysts may demonstrate skills worthy of promotion to Tier 2, trending, or signature management. Give them appropriate room to further tailor data feeds, dashboards, and SIEM content to use cases specific to the site.
- Consider approaches for extending the SOC enclave (described in Section 10) to the remote site for use by the analysts there, perhaps leveraging one of the following approaches:
  - Connect SOC workstations back to the SOC through a strongly authenticated virtual private network (VPN), and ensure that sensitive SOC material is under close physical control.
  - Use a remote thin-client capability with strong authentication if remote site SOC materials cannot be cordoned off from other users.

4.3.5 Centralized SOC with Continuity of Operations

So far, we've discussed scenarios where the SOC has one main ops floor and one place where its management systems and data resides. If the ops floor is taken offline, the CND mission is offline.

Senior constituency leaders and SOC management may decide that some level of physical redundancy is necessary. The purpose, of course, is to ensure continuity of operation (COOP) of CND capabilities in the event of an outage such as the classic "smoking crater" events (thermonuclear war, fire, etc.), weather events (hurricanes, tornados, severe snow, etc.) or power/network outage.

When building a COOP capability for the SOC, there is often an impetus to implement a full-blown "hot/hot" capability whereby a complete duplicate of the SOC's systems (including a second ops floor) is stood up at a location distant from the primary SOC ops floor. This can be very expensive and is not always necessary. Before rushing into a decision for creating a COOP site, the SOC should carefully examine the following decision points:

- What contingencies is the COOP plan designed to address? How realistic are they, and how often are they likely to occur?
- If the main site constituency systems or SOC enclave were "hacked," are the COOP SOC systems designed to be insulated from compromise?
- In the contingencies described, if the SOC was taken out along with the rest of the site where it is located, what constituency systems are left to defend?
- If activation of the SOC's COOP capability were called for and there were any impediments in the process of executing the COOP, would the CND mission actually be a priority in the eyes of constituency seniors?
- Is a full, second instantiation of the SOC warranted?
  - Will a partial duplicate suffice?
  - Does the creation of a secondary COOP site, even if only partial, outweigh other competing resource needs such as more sensors or more personnel?
  - Does the secondary COOP site need to be regularly staffed? If so, should it be for the same hours as the main SOC (such as 24x7) or will regular business hours suffice (8x5 or 12x5)?
- In a COOP scenario, how long can the SOC be down? How quickly must the secondary capability be brought fully online?
- For the COOP site(s) under consideration, does their functionality (such as WAN and Internet connectivity) depend on infrastructure at the SOC's main site? If so, it may be a poor choice.

Many COOP SOC capabilities are built for the classic "smoking crater" scenario, which is very unlikely to occur. COOP is exercised far more often for non-extreme reasons such as network outages, power outages, or major weather events. The other major reason for a SOC to create a second site is essentially to create a second ops floor that can focus on assets at another major site or region of the constituency. In this scenario, we have analysts manning consoles at both locations on a sustained basis, with the analysis workload load-balanced between the two ops floors—perhaps by network/enclave, geographic region, or line of business. This approach is especially handy for constituencies located primarily at two major sites.

Even for SOCs that have a hot/hot COOP capability with servers and analysts at both sites, it is rare that every section of the SOC resides at both locations. More often, we have redundant systems such as the SIEM and IDS management servers, local IDS sensors, Tier 1 analysts, and, perhaps, a couple of sysadmins at the COOP site. In this scenario, it's much easier to coordinate operations between sites than if we also spread Tier 2, trending, intel fusion, sensor management, engineering, and all SOC capabilities between two places.

Regardless of what functions reside at the secondary site, the SOC CONOPS should carefully integrate compensating controls to keep both sites in sync. It also helps to have a lead for the secondary site to coordinate operations with the main site leads and to provide care and feeding for the local analysts. One strategy that may work for SOCs with a hot COOP site in a different time zone is to either match or stagger shifts. By staggering shift changes for the two sites, there is always someone watching the console. For instance, if the main site is an 8x5 operation, the working hours for the secondary site two time zones away could be shifted by an additional two hours, giving four hours of overlap. By doing this, each site is up for eight hours, but together they provide 12x5 coverage.

Creating a second ops floor is very expensive and can be seen as a major drain on resources, especially if regularly staffed. If a SOC wishes to have a secondary COOP "lukewarm" site that it doesn't staff every week, it may consider the following strategy:

1. Choose an existing constituency office building or data center with at least a few spare racks and cubicles.
2. Deploy a redundant instance of key SOC systems such as SIEM, PCAP stores, and IDS management systems, thereby providing failover capability.
3. Find a good spot to place some SOC workstations, perhaps near the TA's office or cubicle.
4. Ensure all security data feeds are directed to both sites or mirrored from the primary to the secondary, at all times.
   a. If the primary site goes offline, having the log data immediately available at the secondary location could be invaluable.
   b. When performing COOP, the amount of time to bring the secondary site online should be minimized. If monitoring systems there are online but not being used, transition is that much quicker.
5. Regularly check (perhaps on a monthly basis) to ensure COOP servers and systems are functional and up-to-date with patches and configuration changes.
6. Schedule semiannual practices of the SOC COOP.

Having redundant core SOC systems will often come in handy. By placing them at a secondary site, the SOC's mission gains an added measure of redundancy. The biggest downside of this strategy is that any touch labor to site systems will come at the expense of the TA's time or sysadmins' travel dollars.

In summary, there are two key elements to an effective SOC COOP capability: (1) create and maintain it against a concrete set of business requirements, and (2) carefully manage the expectations of constituency leadership in the level of continuity the SOC is able to provide.

4.3.6 Centralized SOC with Follow the Sun

In the "follow the sun" model we have three ops floors, each separated by roughly eight time zones. Each ops floor is on the watch during local business hours (e.g., 9 a.m. to 5 p.m.). At 5 p.m. local time, one ops floor rolls to the next ops floor, where it is 9 a.m. This pattern continues every eight hours, giving 24x7 coverage but without making people come to work in the middle of the night.

This approach is very common for IT help desks that serve wide geographic regions (e.g., with major IT vendors and very large corporations). Another advantage is that the operators on shift are more likely to speak the language of those calling during their shift. In terms of pure labor costs, it also may be more affordable than a single ops floor staffed 24x7 because paying people during normal business hours is usually less expensive than paying them to come in at night. However, follow the sun is far less common in security operations because a couple of key assumptions do not carry over.

First, help desks spend a lot of their time talking to users. SOCs certainly interact with users, but they spend most of their time collaborating internally. Therefore, it's important to have all sections of the SOC not only in the same place but at work at the same time. In addition, language barriers and cultural differences among ops floors may be a challenge if each ops floor is staffed by personnel of different nationalities.

Second, although help desks are certainly dynamic environments, SOCs are subject to much more continual change in TTPs. Over the course of a few years, the SOC may completely evolve the way it does business, in response to growing mission demands or new threats. Moving three separate ops floors in the same direction at the same speed is an added challenge.

Third, each SOC Tier 2 and trending analyst will work a number of threads for several hours or days. Handing off an incident from one analyst to the next every eight hours isn't feasible. Either this limits the follow the sun scenario to just Tier 1, or we have three independent Tier 2s, each pursing its respective set of incidents.

If SOC managers wish to pursue a follow the sun approach, it is best to weigh the financial and procedural burdens against the virtues of this model. Of particular importance is the need to synchronize operations and promote cradle-to-grave ownership of incidents.

4.3.7 Tiered SOC

We have introduced the concept of a tiered CND architecture where multiple SOCs operate in a federated manner within a large organization. There are many examples where such an arrangement might be appropriate: within each branch of the DoD, Department of Treasury, Department of Justice (DoJ), Department of Homeland Security (DHS), and so forth. Although each of these entities has one SOC with purview over the entire department, there exist several subordinate SOCs beneath each that perform the majority of CND "heavy lifting" for the organization. These include regional NOSCs under the U.S. Army, Financial Management Service under Treasury, Drug Enforcement Administration under DoJ, and Immigration and Customs Enforcement under DHS. In all of these cases, we have a department- or branch-level SOC, as well as multiple SOCs beneath each.

Both the central and subordinate SOCs have a meaningful role to play, even though those roles can be quite different. Going back to a previous point, we must balance the need to maintain strategic SA with the need to be close to mission assets. Most people familiar with CND are used to operating down in the weeds. This can become a source of conflict in a tiered SOC scenario.

> **In a tiered scenario, the job of the central SOC is to enable security operations across the constituency and maintain a strategic perspective.**

How do we differentiate these roles? Once again, leveraging our capability templates from Section 4.1.2, let's focus on how these two SOCs interact and share the CND mission. (See Table 6.)

It's also important to recognize that not all coordinating and subordinate SOCs fall cleanly into these roles. Some SOCs that sit within a large constituency can support better resourcing, more advanced capabilities, and more strategic reach. Larger organizations can afford more capabilities and, thus, have the potential for greater independence, even though they fall underneath a coordinating SOC. The constructs presented here are only a starting point for establishing roles among tiered SOCs.

Having sorted out the roles and responsibilities of the central and subordinate SOCs, let's look at likely data flows between them. (See Figure 15.)

There are a couple of themes that should emerge here. First, the coordinating SOC handles tasks that scale well across the constituency and can be done in one place. For instance, their expertise in advanced tools and adversary TTPs makes them a good place to formulate training programs for the subordinate SOCs. It's also a great place to perform adversary tradecraft and TTP analysis, because they should have the analysts, the tools,

Table 6. Differences in Roles for Tiered Approach

Responsibility	Central SOC Role	Subordinate SOC Role
Location	Located at or near constituency headquarters	Located at office or headquarters of subordinate constituency
Monitoring, Incident Detection	Across constituency assets not covered by subordinates, such as Internet gateways	Within assigned constituency
Cyber Situational Awareness	Strategic across entire enterprise	Tactical within own constituency
Threat Analysis and Cyber Intel	Strategic across enterprise, reporting to subordinates, detailed analysis of adversary TTPs	Tactical within constituency, consumer of central threat analysis, focused on individual incidents
Incident Response	Cross-constituency coordination, operational direction	Intra-constituency response
Security-Relevant Data Management	Receives summary information and incident reports from subordinates; analysis and retention of data from assets not covered by subordinates, such as Internet gateways	Analysis and retention of own data, augmented with data from other organizations
Training	Coherent program for all analysts in constituency	Execution of general and specialized training for own SOC
Reports to	Constituency executives, external organizations	Own constituency executives, central SOC
Monitoring Capabilities and Tools	Enterprise licensing, lead on tool deployment and refresh	Chooses monitoring placement, specialized gear when needed

the time, and just enough knowledge of constituency networks to make sense of the artifacts handed to them by subordinate SOCs.

Second, it's the job of the subordinate SOCs to perform most of the tactical hands-on monitoring, analysis, and response to incidents. The coordinating SOC is there to make sure its entire constituency is working toward a common goal, and that they have shared SA. While the subordinate SOCs may provide a limited event stream to the coordinating SOC, it's unlikely the coordinating SOC analysts have the context to make sense of that data. Incident reporting and trending from the subordinate SOCs support coherent SA formulated by the coordinating SOC.

Third, it is more likely that the central coordinating SOC will have a sizable budget for big technology purchases and custom tool development. Requiring subordinate SOCs to use

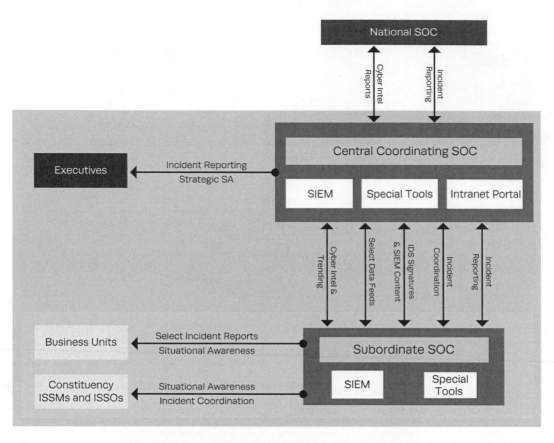

Figure 15. Data Flows Between Central and Subordinate SOCs

a specific product may be too heavy-handed. Instead, what may work is to mandate the use of a type of tool and provide free copies of one specific brand. As a result, the subordinate SOCs can use the enterprise tool if it fits their needs or pay for their own if it doesn't.

Last, and perhaps most important, the coordinating SOC must work very hard to maintain relevance and usefulness in the eyes of the subordinate SOCs. The SOCs at the bottom of the food chain are typically sitting on a pile of raw data. The coordinating SOC has little apart from the incident reporting, cyber intel, and select data feeds from its subordinates. The coordinating SOC must also be careful that downward-directed guidance and tasking are perceived as relevant and useful.

They must work in a symbiotic relationship that stems from perceived value and analyst-to-analyst contact, far more than mandate and policy. The coordinating SOC may offer substantial help in the form of in-depth forensics capabilities, cyber intel, and SA to its subordinates, in exchange for the subordinates' processed incident reports. The subordinates turn data into information; the coordinating SOC turns information into knowledge. This relationship is self-reinforcing over time and, usually, must begin by the coordinating SOC offering something of value to its subordinates that these subordinates cannot get on their own, such as tools and authority.

4.3.8 Coordinating SOCs

At the most extreme end of constituency size are national coordinating SOCs, which we will colloquially refer to as "mega-SOCs." Today, there is limited agreement on the proper role of these organizations. They are comparatively few in number so their capability portfolio and influence are subject of some debate. National-level coordinating SOCs have a unique mission; their goals include:

- Forming a coherent SA picture for their entire constituency, focusing on constituency vulnerability to threats, and adversary TTPs
- Harmonizing operations among their subordinate SOCs
- Bringing their subordinate SOCs up to a baseline set of capabilities.

By contrast, most CND analysts and leaders are used to operating down in the weeds where they have access to raw data, have some measure of vested authority over their constituency, and are direct participants in incident response. The "mega-SOC" doesn't always have these things, and when it does, they often take on different forms than with the mega-SOC's subordinates.

Instead of focusing on direct reporting of raw event feeds or promulgating detailed operational directives, the coordinating SOC may achieve its goals by providing a unique set of capabilities that its subordinates usually can't. These include:

- Providing secure forums for collaboration between subordinate SOCs (e.g., wikis and secure online forums)
- Performing strategic analysis on adversary TTPs by leveraging a wealth of finished incident reporting. A mega-SOC is uniquely positioned to focus on observing and trending the activity of key actors in the cyber realm.
- Providing a clearinghouse of tippers, IDS signatures, and SIEM content that other SOCs can directly leverage without further legwork. A mega-SOC could harvest indicators from human-readable cyber intelligence and provide it back out in both human- and machine-readable form for ingest by subordinates' analysts and SIEM, respectively. In order for this to work, however, intel should be turned around in a timescale and with detail that is beneficial to its recipients. This will likely mean processing and redistributing cyber intel in timeframes of hours or perhaps a few days, and in so doing preserving as much original detail and attribution as possible.

- Aggregating and sharing CND best practices, process documents, and technical guidance
- Providing malware analysis and forensic services to constituent SOCs that have collected the necessary files or images but don't have the staffing to analyze them
  - Of all the things a mega-SOC can do, this is potentially one of the most impactful, because many SOCs have a hard time maintaining the skill set to perform the malware analysis or forensics that is critical to have during a major incident.
  - This can include an automated Web-based malware detonation "drop box" (See Section 8.2.7.) or in-depth human analysis of media or hard-drive images.
- Providing enterprise licensing on key CND technologies such as network and host monitoring tools, vulnerability scanners, network mapping tools, and SIEM, provided the following two conditions are met: (1) subordinates are not forced to use a specific product, and (2) there is sufficient demand from subordinates to warrant an enterprise license
- Providing CND analyst training services:
  - On popular commercial and open source tools such as IDS and malware analysis
  - On the incident response process
  - On vulnerability assessment and penetration testing
  - Leveraging a virtual "cyber range" where analysts can take turns running offense and defense on an isolated network built for Red Team/Blue Team operations
  - Running SOC analysts through practice intrusion scenarios, using real tools to analyze realistic intrusion data.

In many ways, these services are far less glamorous than flying big sensor fleets or collecting large amounts of raw data, especially to those running the mega-SOC. From the perspective of the constituent SOCs, they are far more valuable, *because they provide something back.* By providing these services, the Mega SOC is likely to achieve its unique goals better than if it tries to provide the same capabilities as its subordinates.

For more details on standing up a national SOC, see [60].

Chapter 5

Strategy 3: Give the SOC the Authority to Do Its Job

The SOC must execute its mission against constituency IT assets that almost always belong to someone else. Even though the SOC is usually a member of the constituency it serves, primary ownership and operation of hosts and networks is vested in another member of the constituency that the SOC must interact with. As a result, the SOC's ability to assert proactive and reactive authority must either be codified through written authority or inherited through the SOC's parent organization. In our third strategy, we will address both of these issues in turn: (1) what authorities the SOC needs and (2) what organizational alignment will best aid the CND mission.

5.1 Written Authorities

Every SOC exists based on some sort of written policy that grants its authority to exist, procure resources, and enact change. In addition, there are a host of supporting IT and cybersecurity policies that a constituency should have that enable a SOC to execute its mission. In this section, we will cover both in order. When crafting policy, you may want to consider the additional guidance in Section 2.5 of [8] and the policy templates found in [61].

5.1.1 Charter Authority

SOCs are generally considered effective if they have the ability to both detect incidents and direct countermeasures. Bringing these roles together requires a well-written charter and supporting cybersecurity policies. While the SOC must get along with various cybersecurity stakeholders, the SOC must frequently refer to formal doctrine.

> **An effective SOC should have a charter and set of authorities signed by constituency executive(s) in order to press for the resources and cooperation needed to execute its mission.**

This charter may also include the services expected of the SOC, but should not inhibit a SOC from growing into its roles as its maturity and resources progress. Even in cases where a SOC has not fulfilled the entire scope of its charter, having that piece of paper helps the SOC grow into such a role. The charter describes what a SOC *should* be doing, not just what it *is currently capable of* doing. It is important also to recognize that the charter does not describe *how* a SOC fulfills its mission, only *what* it does and *who* has supporting responsibilities.

A charter also helps eliminate misconceptions about what the SOC is and what it must do. Conversely, SOCs that lack such written authority often have to spend a lot more energy begging for help and less time making a positive impact.

Every organization has a different approach to writing IT/cybersecurity policy. With that in mind, we will consider the "whats" of policy and authorities that will enable a SOC to execute its mission, without focusing too much on the "hows." How these elements are allocated among a charter and other policy documents may vary. The main distinction is that the core scope of the mission should always get the signature of the constituency's chief executive. Other items may be codified elsewhere and, therefore, updated with greater frequency.

The following policies are written for a tiered SOC model. Readers can take this template and modify it according to their own organizational model and offered capabilities.

5.1.2 Lowest Tier SOC

The following elements should be codified in the charter of a SOC that is the sole CND provider for a given constituency or that sits at the lowest tier in a multitiered arrangement:

- To function as the operational center and head of cyber intrusion monitoring, defense, and incident response for the constituency
- Within its constituency, the authorities to:
  - Deploy, operate, and maintain active and passive monitoring capabilities, both on the network and on end hosts

- Proactively and reactively scan hosts and networks for network mapping, security configuration, and vulnerability/patch status
- With SOC line manager approval, coordinate or directly apply countermeasures (including the termination of network resources) against systems and user accounts involved in an incident, in coordination with system and network owners
- Respond directly to confirmed incidents, in cooperation with appropriate parties, possibly including reporting outside of their management chain to entities such as the CIO or CEO, if necessary
- Gather, retain, and analyze artifacts such as audit log data, media images (hard drive, removable media, etc.), and traffic captures from constituency IT system to facilitate incident analysis on both an ad hoc and sustained basis, recognizing any handling caveats derived from applicable laws, regulations, policies, or statutes.
- To be recognized by help desk staff, ISSMs, ISSOs, and IT support staff when reporting, diagnosing, analyzing, or responding to misconfiguration issues, outages, incidents, or other problems that the SOC needs external support to resolve
- To architect, acquire, engineer, integrate, operate, and maintain monitoring systems and the SOC enclave
- To exercise control over funding for tool engineering, maintenance, staffing, and operational costs
- To support any other capabilities it intends to offer, such as security awareness building, cybersecurity education/training, audit collection, or insider threat monitoring.

5.1.3 Central Coordination SOC

If the SOC follows a tiered model, the central coordinating SOC will likely need the following authorities in addition to those of the lowest tier SOC:

- Serve as an entity that is operationally superior to subordinate SOCs
  - Gather security-relevant data from all subordinate SOCs
  - Coordinate response actions among subordinate SOCs
  - Direct improvements to subordinate SOC capabilities and operations, in accordance with fulfilling incident response requirements across the greater constituency.
- Manage devices that aggregate security-relevant data from subordinate SOCs and sensors directly placed on hosts and networks
- Act as the focal point for enterprise-wide security information sharing and SA through common practices, SOC-provided/developed tools, and preferred technologies or standards
- Propose enterprise-wide *preferred* standard network and security monitoring technologies and practices

- Negotiate enterprise-wide licensing/pricing agreements of monitoring technologies that may benefit subordinate SOCs, where possible.

5.1.4 Other Enabling Policies

Apart from the policies that directly enable a SOC to function, there are a number of other IT and cybersecurity policies that enable effective security operations. The SOC should consider influencing or providing comment on these policies or seeing that they are created, if they do not already exist:

- **User consent to monitoring:** Giving the SOC and auditors the unambiguous ability to monitor and retain any and all activity on all systems and networks in the constituency
- **Acceptable use policy:** IT system usage rules of behavior, including restrictions on Internet and social media website use and authorized software on constituency systems
- **Privacy and sensitive data handling policies:** Instructions for managing and protecting the types of information flowing across the monitored network, including personal, health, financial, and national security information
- **Internally permitted ports and protocols:** Enumeration of the ports and protocols allowed within the constituency, across the core, and through enclave boundaries
- **Externally permitted ports and protocols:** Enumeration of ports and protocols allowed by devices through external boundaries such as through a demilitarized zone (DMZ), to business partners and to the Internet
- **Host naming conventions:** Describing conventions for naming and understanding the basic type and role of IT assets on the basis of their DNS record
- **Other IT configuration and compliance policy:** Everything from password complexity to how systems should be hardened and configured
- **Bring your own device and mobile policies (if applicable):** Rules that govern how employees may interface with constituency networks, applications, and data with personally owned IT equipment and mobile devices
- **Approved OSes, applications, and system images:** The general approved list of OSes, applications, and system baselines for hosts of each type—desktops, laptops, servers, routers/switches, and appliances
- **Authorized third-party scanning:** Rules for notifying the SOC when another organization wishes to perform scanning activity such as for vulnerabilities or network discovery
- **Audit policy:** High-level description of the event types that must be captured on which system types, how long the data must be retained, who is responsible for

reviewing the data, and who is responsible for collecting and retaining the data—with recognition of the performance impact value of the data gathered

- **Roles and responsibilities of other organizations with respect to incident response:** Most notably those bolded in Appendix A
- **Written service level agreements (SLAs) where applicable:**
  - Network capacity and availability requirements
  - Contingency planning if contracted network services fail
  - Network outage (incident) alerts and restoration and escalation/reporting times
  - Security incident alerts and remediation procedures and escalation/reporting times
  - Clear understanding of each party's responsibilities for implementing, operating, and maintaining the security controls or mechanisms that must be applied to the network services being purchased.
- **Legal policies:** Concerning classifications of information, privacy, information retention, evidence admissibility, and testifying during investigations and prosecutions of incidents.

5.2 Inherited Authorities

The SOC draws its authorities, budget, and mission focus from the organization to which it belongs. Making the right choices about where to put the SOC in the organization chart can propel it to greater success than it would otherwise be capable of. For most SOCs, their placement within the constituency is a function of decisions that were made when the SOC was first formed. During either creation or corporate restructuring, there are opportunities for executives to fine-tu ne the SOC's organizational placement.

The SOC's organizational placement is keenly influenced by the following factors:

- Organizational depth—how far down in the organization chart the SOC is placed. Does it report directly to the CIO or is it buried 15 levels deep? If the latter, what policy and process can be put in place to mitigate this?
- What authorities the parent organization has—both on paper and in practice.
- The power and influence wielded by parent organization executives. Are they attuned to the CND mission, can they support authority models discussed in Section 2.3.3, and are they likely to stay "hands off" from day-to-day security operations?
- Established funding lines and budget of the parent organization. Are they able to fund tools for comprehensive monitoring and people who can staff all the capabilities implied by the SOC's charter?
- What capabilities the SOC will offer.
- What organizational model the SOC features. If it is tiered, can the subordinate "mini-SOCs" live within business units while the coordinating SOC sits under the

CIO? Or, if it is a centralized SOC, can it sit near the NOC and preside over the entire constituency?

The choice of organization placement is intertwined with another, perhaps more interesting question: "Who is in control of defending the enterprise?" More often than not, there are multiple executives who feel they are the protectors of the mission. Under which executive would the SOC flourish, and how far down the command chain should it sit?

In order to direct defense of the enterprise, we need two things: (1) SA over the constituency, down to specific incidents and the systems and mission they impact, and (2) the authority and capability to direct changes to IT systems proactively or in response to an incident—such as changing domain policies, routers, or firewalls or pulling systems off the network.

As we discussed in Section 4.1.1, these two needs can be at odds with each other, creating tension in where the SOC s hould sit and to whom it should report. Several executives—CEO, COO, chief security officer (CSO), CIO, CISO, CTO, and their subordinates—have at least some sense of ownership over cybersecurity and have a legitimate need for the SA a SOC can provide. When there is a serious incident, it is likely that many of these parties will want to be informed. Things get sticky when multiple parties assert (potentially conflicting) roles in directing, implementing, or approving response to an incident. A SOC's actions are often second-guessed, especially during an incident. Gaining "street cred" through a well-advertised track record of handling incidents competently helps offset this. But this is gained slowly over many good deeds and lost easily by few mistakes.

Clarifying who truly gets to call the shots through policy signed by the chief executive of the constituency is critical. This is true regardless of the authorities delegated to the SOC. Each one of these executives serves as a candidate for the management a SOC will serve under and, thus, its organizational placement. This can occur intentionally through charter, by the demands of the larger mission or business needs, or simply by accident of where a SOC was first formed.

Despite all of this, the SOC is distinct and separate from almost any other part of the constituency, even though it may be near a NOC, CISO, or security function. Its skills, attitude, mission, and authorities always set it apart. As a result, the following is almost always true:

> **Regardless of organizational placement, a SOC often feels like the ugly duckling. This compels the SOC to regularly connect with partner organizations and clearly articulate its role and value.**

Regardless of where we place the SOC, it must have budgetary, logistical, and engineering support in place to serve its sustained operations and growth. It must not be broken into disparate pieces. In the following subsections, we briefly discuss some pros and cons of some popular organizational alignments.

5.2.1 Subordinate to the Chief Information Officer or the Chief Information Security Officer

This arrangement makes the most sense with large constituencies where IT operations and the NOC fall under the CIO or CISO. They have an "ops" slant while maintaining strategic visibility and authority. If this is not the case and the CIO is mostly oriented toward IT policy and compliance, this can be a bad arrangement for the SOC. It most likely will lack "street cred" and will not maintain a CND focus. Sometimes a deputy CISO or deputy CIO position may be created whose sole responsibility is to manage the SOC (preferably given great leeway by the CIO). A SOC organized under a CIO or CISO who can support a true ops tempo with tactical visibility and connections can work very well. In many cases, where the SOC is organized somewhere *other* than the CISO, it will have some sort of "dotted line" relationship whereby the CISO or CIO influences SOC actions and focus.

5.2.2 Subordinate to the Chief Operations Officer

This can be a positive arrangement for the SOC, assuming operations functions of the constituency are consolidated under the COO. In such a scenario, the SOC is more directly involved in meaningful conversations about the daily operations and mission of the constituency as a whole. If there is a daily ops "stand-up," the SOC may have direct representation. It is also more likely that the SOC's needs will be met through adequate policy, budget, and authorities. This arrangement can be looked upon fondly by some SOCs because of its visibility, but the SOC must be careful what issues it brings to the COO's desk, for fear of "crying wolf."

On the downside, it can be a challenging position because the SOC will likely compete for the COO's limited time and money. If the COO does not have direct, meaningful control over constituency operations or the COO's function is seen as "overhead," the SOC can inherit this reputation as overhead and be vulnerable to cuts during budget time.

5.2.3 Subordinate to the Chief Security Officer

A SOC almost always leans on information security professionals located across an enterprise to help establish visibility and support response at remote sites. Alignment under security can help strengthen this. It can also help if these security bodies are able to seamlessly take care of IT compliance and misuse cases. Doing so (as is the case with many ISSOs in government) leaves the SOC to focus on more advanced cyber threats. However, constituents' potentially negative perception of "security," "overhead," or "compliance" functions may not help. In addition, it may sometimes be too much of a stretch for an organization responsible for physical protection to take on a large portion of the cybersecurity mission.

The biggest challenges to a SOC organized under a security function are (1) maintaining strategic perspective and partnership with the CIO or CISO while having day-to-day visibility and communication with IT and network ops, (2) separating its function from IT and security compliance, and (3) ensuring the right mix of technical expertise. Again, the SOC must be careful, from a budgetary perspective, because security functions are often seen as overhead. Also, many organizations do not have a separate security organization apart from the CIO and CISO, ruling this out as a possibility.

5.2.4 Peered with the Network Operations Center Under IT Operations

The most common organizational placement of a SOC is having it collocated logically or physically, or both, next to or as part of a NOC. This provides a number of obvious virtues: 24x7 operations are merged, and response actions can be swiftly adjudicated through a single authority that balances the real-time needs of security and availability. Furthermore, there are many devices such as firewalls and IPSes that are seen as shared security and network capabilities. Consolidating both functions onto one watch floor (with distinct staff and tools) will save money, especially for enterprises that can't justify having a separate SOC.

Keeping network and security ops as distinct *peers* with separate people, tools, and funding will help avoid sidelining security in the name of network availability.

> **In order to support a healthy constituency, NOC and SOC functions should be viewed as equal partners, rather than one subordinate to the other.**

Even though NOC and SOC roles should be clearly divided, seamless collaboration is key, whether or not they are collocated. If network and security ops are collocated, it is a good idea to coordinate shift patterns and other staffing logistics. A SOC will always field some calls from users. However, directing *some* of these calls to a nearby help desk (also under IT ops) could help reduce the call load. Similarly, other parts of IT ops can perform network scanning and patch management, something a SOC should watch closely. Finally, with both, network and security ops can report to one management authority that (it is hoped) has unambiguous authority to direct changes to the network for both availability and security reasons. *This alone is reason enough to place the CND mission inside IT ops.*

SOCs that are able to leverage these strengths while maintaining strategic perspective—through partnerships with executives such as CIOs, CISOs, and CEOs—often have a great chance of succeeding.

5.2.5 Embedded Inside a Specific Mission or Business Unit

Pigeonholing CND ops within a given business area may severely limit visibility to what is within that business unit. But it often presents unique opportunities for the SOC to be mission-oriented in how it monitors and responds. If a particular portion of an enterprise's mission is very sensitive, having a SOC just for that mission can help. However, these efforts must be tied back to an enterprise-wide visibility and coordination capability (such as a tiered SOC model). Alternatively, a distributed SOC model with representatives in each business unit and the main SOC viewed as a headquarters function may work.

Chapter 6
STRATEGY 4: Do a Few Things Well

··

There are only a few prerequisites for being a SOC: the ability to (1) accept incident reports from constituents, (2) help constituents with incident response, and (3) communicate information back about those incidents. This represents only a fraction of the SOC's potential duties. The question is, what other capabilities should it provide? In our fourth strategy, we explore the possibilities for what capabilities a SOC may offer. Our primary objective is to limit capability "sprawl" so we can focus on doing a few things well rather than many poorly. Going along with this, we want to (1) carefully manage expectations of constituency members and executives, (2) enhance trust and credibility with the constituency by handling each incident with care and professionalism, (3) avoid stretching limited SOC resources too thin, and (4) take on additional roles or tasks only when resources, maturity, and mission focus permit.

6.1 Introduction

Before we decide which capabilities to provide, we must ask ourselves some critical questions:

- What is the intended scope of the SOC mission? Do evolving needs of the constituency compel us to alter or expand the SOC mission?
- Is it appropriate for the SOC to engage in direct monitoring and response activities, or is it more productive for the SOC to coordinate and harmonize the activities of other SOCs?

- How does the SOC's organizational placement bias its focus? For instance, is it relied upon to provide SA to constituency executives, or perhaps to implement rapid counter-measures? Can the SOC balance these obligations against other mission priorities?
- What capabilities exist elsewhere that the SOC can call on when needed, such as audit retention and review, vulnerability assessment and penetration testing, artifact analysis and malware analysis, or countermeasure implementation? Are any of these performed by an organizationally superior coordinating SOC?
- Does SOC resourcing enable it to reach beyond firefighting to advanced capabilities such as tradecraft analysis, custom signature creation, or tool research and development?
- If the SOC offers a given capability, will it be exercised enough to justify the associated costs?

The SOC will always share control over the scope of its mission with external forces such as edict and policy handed down by constituency executives. Moreover, careful attention must be paid to the perceived expectations of the constituents, versus what capabilities the SOC is in a position to actually support. Taking another cue from Section 2.3.3 of [8], we emphasize *quality* of capabilities offered versus *quantity*—do a few things well rather than a lot of things poorly. In a world where the SOC must always work with external entities to complete its mission, "street cred" is paramount.

6.2 Capability Templates

In Table 7, we illustrate a *typical* capability offering for each of the five SOC templates we described in Section 4.1.2, using the following descriptors:

- **B: Basic.** The SOC has a partial or basic capability in this area.
- **A: Advanced.** The SOC has a well-developed, mature capability in this area.
- **O: Optional.** The SOC may or may not offer this capability, potentially deferring it to an external group either within the constituency or from another SOC, if appropriate.
- **–: Not Recommended.** Usually due to constituency size or distance from end assets and mission, it is not recommended that the SOC support this capability.

It is important to recognize that this table describes the capabilities of each SOC *once they have matured into a steady state*. In other words, it outlines a target state, not a maturation path; we address growth and dependencies in Section 6.3. Additionally, this table serves as a starting point for SOCs wishing to pick from a menu of capabilities—they must always tailor what they take on and how they fulfill those needs.

Table 7. Service Templates

Name	Virtual	Small	Large	Tiered	National
Real–Time Analysis					
Call Center	O	B	A	A	A
Real-Time Monitoring and Triage	O	B	A	A	O
Intel and Trending					
Cyber Intel Collection and Analysis	B	B	A	A	A
Cyber Intel Distribution	O	B	A	A	A
Cyber Intel Creation	–	O	B	A	A
Cyber Intel Fusion	O	B	A	A	A
Trending	O	O	A	A	A
Threat Assessment	–	O	B	B	A
Incident Analysis and Response					
Incident Analysis	B	B	A	A	O
Tradecraft Analysis	–	O	A	A	O
Incident Response Coordination	B	B	A	A	A
Countermeasure Implementation	O	O	O	O	O
On-site Incident Response	B	O	O	O	O
Remote Incident Response	B	B	A	A	O
Artifact Analysis					
Forensic Artifact Handling	O	B	A	A	O
Malware and Implant Analysis	–	O	B	A	O
Forensic Artifact Analysis	–	O	B	A	O
SOC Tool Life–Cycle Support					
Border Protection Device O&M	O	O	O	O	O
SOC Infrastructure O&M	O	B	A	A	A
Sensor Tuning and Maintenance	B	B	A	A	O
Custom Signature Creation	O	O	A	A	A
Tool Engineering and Deployment	O	B	A	A	A
Tool Research and Development	–	–	O	B	A

Table 7. Service Templates

Name	Virtual	Small	Large	Tiered	National
Audit and Insider Threat					
Audit Data Collection and Storage	O	O	B	O	–
Audit Content Creation and Management	O	O	O	O	–
Insider Threat Case Support	B	B	A	A	–
Insider Threat Case Investigation	–	O	O	O	–
Scanning and Assessment					
Network Mapping	O	O	B	B	O
Vulnerability Scanning	O	B	A	O	–
Vulnerability Assessment	O	O	O	O	O
Penetration Testing	O	O	O	O	O
Outreach					
Product Assessment	O	O	O	O	O
Security Consulting	O	O	O	O	O
Training and Awareness Building	O	O	O	O	O
Situational Awareness	O	B	A	A	A
Redistribution of TTPs	–	O	B	A	A
Media Relations	–	–	O	O	B

Rather than consuming many pages on exhaustive justification for each of these choices, we offer the following discussion on some major decision points:

- **Call center.** Small SOCs may transfer this function to either the help desk or the NOC. However, this may be problematic because the CND implications of a network outage or help desk call may not be obvious to someone not trained in cyber incident analysis and response. The SOC does not want to miss tips that otherwise would have been handled just as an availability issue. In either event, it helps to give the help desk clear escalation criteria and paths to the SOC. One other option for non-24x7 operations is to direct the SOC hotline to a cell phone or pager, whose assignment is rotated among SOC members on a daily or weekly basis.
- **Real-time monitoring and triage/incident analysis.** This capability is core to virtually all SOCs. In very large coordinating SOCs, this will take on a slightly different

form. Rather than looking at raw data and receiving tips from end users, the SOC will likely receive calls and incident reports from other SOCs.

- **Malware and implant analysis/forensic artifact analysis.** These capabilities are hard to staff in small and medium-sized SOCs for two reasons: (1) there needs to be a steady stream of incidents requiring in-depth postmortem, attribution, or legal action, and (2) personnel qualified to do this are in high demand. It is often an effective strategy for a smaller SOC to rely on a coordinating SOC or service provider to provide this on an as-needed basis. For SOCs that handle more than a few major cases a week, keeping at least a basic media forensics capability in house will speed response and recovery efforts.
- **On-site incident response.** This is a function of whether a SOC's organizational model incorporates elements of the internal distributed SOC model and how geographically dispersed the constituency is.
- **Sensor tuning.** If a SOC has its own fleet of monitoring equipment (e.g., network or host IDS/IPS), tuning must be a sustained, internal activity driven by feedback from analysts. Therefore, *this capability is considered a requirement for any SOC that deploys monitoring capabilities*. It is also important to recognize that sensor tuning is a continuous process necessary for the correct functioning of the sensor fleet. It is, therefore, an operations function and not an engineering or development function.
- **Tool engineering and deployment.** Whether a SOC performs this in house or depends on an external organization, *this capability must exist somewhere in the larger organization to which the SOC belongs*.
- **Vulnerability assessment and penetration testing.** The SOC is a natural place to house and coordinate these activities. It provides a unique basis for enhanced SA and raises visibility of the SOC as a resource to system owners and security staff. Because the staffing needs for Blue and Red Teaming can become quite large, some SOCs may choose to eschew the resource demands it entails. They thereby avoid the perception of being a large cost center. Larger SOCs are a more likely home for vulnerability assessment/penetration testing (VA/PT) capabilities. Some smaller constituencies may outsource this to a third party.
- **Network mapping/vulnerability scanning.** These capabilities may be viewed as sustaining IT functions rather than CND or security operations functions. Subsequently, the choice to include these in the SOC should occur on a case-by-case basis. Whether or not these are placed elsewhere, the SOC may be well-advised to aggregate and synthesize scan results, because up-to-date network maps and vulnerability statistics are key inputs to incident analysis efforts.

- **Border protection device O&M/countermeasure implementation.** These capabilities are rarely offered by a SOC that is not organizationally adjacent to, or comingled with, IT/network operations, as in a NOSC. These capabilities are also usually limited to SOCs with full reactive authority. Adjacent security analysis and network management functions enable quick-turn response, as directed by a SOC manager or SOC shift lead.
- **Audit data collection, retention, storage/audit content creation, and management.** These capabilities are likely seen as a side benefit of comprehensive log collection and analysis. Large SOCs that incorporate distributed resources will likely find it necessary to provide a robust audit collection and redistribution capability since forward-deployed SOC personnel will be more focused on local systems and data sources.

Decision making about which capabilities a SOC should offer often draws in the subject of organizational placement. It's usually a given that a certain capability should be provided by *someone*. The question is whether that function finds its home in the SOC or somewhere else.

6.3 Capability Maturation

We have described a target capability template for each of our five SOC templates. In this section, we show some potential growth patterns that SOCs typically take when expanding into new capability areas. We also use this to identify dependencies and relationships among capabilities.

- **Cyber intel collection and analysis/sensor tuning and maintenance to cyber intel fusion/custom signature creation.** Continual exposure to multiple sources of cyber intel will train analysts to be more discriminating in what they gather, and will help them recognize how their own defenses can be enhanced. Knowledge of adversary TTPs, constituency environment, and how to write signatures naturally leads analysts to crafting their own IDS signatures and SIEM analytics.
- **Incident analysis, malware and implant analysis/custom signature creation to cyber intel creation and distribution/redistribution of TTPs.** Over time, analysts' experience with individual incidents should grow into a more macroscopic understanding of adversary behavior. This can lead analysts to draw observations and conclusions they may wish to share with constituents or other SOCs, further reinforcing their SA.
- **Incident analysis to forensic artifact analysis, and/or to malware and implant analysis.** As the volume of incidents increases, so too should the SOC's consistency and efficiency in handling those incidents. Analysts' need to establish root cause analysis often leads them in one or both of the following directions. First, as the volume

and complexity of incidents caught increase, so too will the number of traffic capture and media artifacts. What may start as ad hoc artifact analysis will likely turn into a repeatable, rigorous process involving dedicated forensics specialists and tools. Second, the amount of malware caught will likely increase, and thus the need to understand the comparative threat posed from one incident to the next—is this typical malware or a targeted attack? The SOC may evolve proactive means to regularly extract suspect files from network traffic, and perform static and/or dynamic malware analysis on those files.

- **Cyber intel fusion/incident analysis to trending/tradecraft analysis/threat assessment.** The SOC should recognize that a knee-jerk response to all incidents of reformat and reinstall is not always the best course of action. By engaging with other parties, the SOC should gain the authority and ability to passively observe the adversary and better gauge the extent of an intrusion. This will also compel the SOC to support a more strategic perspective on the adversary, allowing the SOC to perform advanced long-term trending. That, along with a keener understanding of constituency mission, means the SOC can author threat assessments that help guide future monitoring efforts and changes to the constituency's security architecture and major system acquisitions.

- **Tool engineering and deployment to tool research and development.** As mission demands grow, the SOC will likely run into the limits of COTS and FOSS capabilities, leading the SOC to develop its own tools. This will likely start with projects that "glue" multiple open source and commercial capabilities together in new or different ways. In more extreme examples, well-resourced coordinating SOCs may put together, from scratch, polished capability packages for their constituents.

For additional examples of capability maturation, the reader may turn to Section 2.7.5.1 of [4], which served as the inspiration for this section.

Chapter 7
Strategy 5: Favor Staff Quality over Quantity

People are the most important aspect of CND. It's a cliché in virtually all areas of business, but it's true. SOCs have a limited budget with which to compete for a finite pool of talent, making staffing the SOC a challenge. With the right tools, one good analyst can do the job of 100 mediocre ones. As a result, we offer a critical point:

> **Analyst quality is vastly more important than analyst quantity.**

Moreover, while analysts can be trained to use a tool in a rudimentary manner, they cannot be trained in the mind-set or critical thinking skills needed to master the tool. This forms the basis for our fifth strategy: exercise great care in hiring and keeping CND analysts. Choose staff quality over quantity. We break this strategy down into three parts which we cover in order: (1) whom do we hire, (2) ideally, how many staff do we need, and (3) how do we retain them.

7.1 Whom Should I Hire?

In this section, we discuss the traits of the ideal SOC hire. We examine qualities in relation to a candidate's mind-set, background, and skill set, each of which is covered in turn.

7.1.1 Mind–set

Perhaps the number one quality to look for in any potential hires to the SOC is their passion for the job, regardless of the position. Intrusion monitoring and response is not just "a job" where people put in their eight- or 12-hour shift, collect a paycheck, and then leave. When it comes to "cyber," we're looking for enthusiasm, curiosity, and a thirst for knowledge. This passion is what will keep them coming back to the job, day after day, despite the stress and challenges inherent in operations. This passion, along with intellect and other soft skills, is what propels fresh recruits into becoming what we will call "rock-star analysts." Seasoned rock-star analysts can do all or most of the following:

- Pick out potential intrusions from seemingly benign sets of audit logs or IDS alerts
- Build new tools and techniques to compress human-intensive tasks into work that can be achieved in a fraction of the time
- Gather disparate data (e.g., system logs or hard drive images), construct a timeline of events, and evaluate the disposition of a potential intrusion
- Pick up on subtle cues with network protocol analysis tools to recognize the meaning and implications of traffic across all seven layers of the Open Source Interconnection (OSI) network protocol stack
- Tear apart a piece of malware and formulate a working understanding of its attack vector and likely purpose
- Identify system misconfigurations and work with system owners to correct them
- Establish and grow relationships with members of the SOC and partner SOCs, sharing best practices, tools, and tippers
- Put themselves in the shoes of the adversary, look at the structure of a network and supported mission, and assess where there is cause for concern.

Talent attracts talent, and finding just a few rock-star analysts can propel the SOC forward by leaps and bounds. While the entry-level positions in the SOC require repetition and structure, more advanced positions require a different mind-set:

- Strong intuition and ability to think "outside the box"
- Attention to detail while seeing the bigger picture
- Ability to pick up new concepts; thirst for knowledge
- Desire to script and automate repetitive parts of the job.

One strategy for SOC hiring managers is to find people ripe for becoming rock-star analysts and who have the capacity to learn the procedures and tools for doing so.

> **Intrusion analysis cannot and never will be turned into a completely formulaic, repeatable process with every step defined in exhaustive detail.**

Analysts must be free to analyze. It is indeed true that Tier 1 analysts have more structure in their daily routine for how they find and escalate potential intrusions. However, those in upper tiers must spend a lot of their time finding activities that just "don't look right" and figuring out what they really are and what to do about them. *Overburdening analysts with process and procedure will extinguish their ability to identify and evaluate the most damaging intrusions.*

7.1.2 Background

CND analysis requires a superb understanding of how networks and systems operate—in practice, not just in theory. Although rarely expert in *all* areas, analysts can usually answer questions such as:

- How do you install and configure a Linux or Windows system?
- What does normal DNS traffic look like?
- How do you correctly architect a network perimeter DMZ?
- What is wrong with this switch configuration?
- How can I achieve common tasks in popular scripting and programming languages?

An ideal candidate should be able to demonstrate a general "literacy" of IT and cybersecurity, along with deep knowledge in at least one or two areas related to CND—a concept known as the "T-shaped person" [62]. This knowledge is usually gained through a combination of the following three things:

- Formal training in IT, computer science (CS), electrical or computer engineering, cybersecurity, or a related field
- On-the-job experience in IT operations, system/network administration, or software development
- Self-study in system administration, software coding, CND, and vulnerability assessment/penetration testing, often achieved in candidates' spare time

With the growing number of undergraduate and graduate-level programs specifically in information security, forensics, cryptography, and malware analysis, we see a rising number of applicants who specifically tailored their formal education to a career in security operations, either at the undergraduate or graduate level.

That said, making a formal degree or five years of experience in IT a universal requirement for incoming analysts isn't absolutely necessary. Some SOCs focus on assessing candidates' practical IT experience, thirst for knowledge, and ability to think outside the box. Candidates who haven't been in IT very long, but demonstrate solid problem-solving skills, could be initially assigned to Tier 1 and allowed to grow into more advanced roles as their skills advance.

There are several previous positions that candidates may come from: help desk, ISSO or ISSE, IT/cybersecurity policy writing and compliance, software and systems development, and system administration. In any of these cases, it is important to assess candidates' breadth and depth of technical knowledge, ability to assimilate and use new information, and appreciation for the realities of IT operations.

7.1.3 Skill Set

Mature SOCs should have a robust training program that brings new recruits up to speed on the TTPs the SOC uses to execute its mission. Candidates with a background in either system administration or penetration testing can usually pick up the CND specifics in a matter of weeks or a few months. Although it's easy to focus on experience with various CND tools, as a job qualification this is only one-third of the picture. Many experienced SOC leaders would argue that understanding how a tool works and what it tells the analyst is more important than the semantics of how one specific tool works.

Going back to the concept of the "T-shaped person," a seasoned CND analyst should be able to demonstrate general knowledge of most of the following, with deep understanding in at least one or two areas:

- Linux/UNIX system administration, along with network (router and switch), Web server, firewall, or DNS administration
- Work with various FOSS IDS/IPS, NetFlow, and protocol collection and analysis tools such as Snort [63], Suricata [64], Bro [65], Argus [66], SiLK [67], tcpdump [68], and WireShark [69]
- Working knowledge of entire TCP/IP or OSI network protocol stack, including major protocols such as IP, Internet Control Message Protocol (ICMP), TCP, User Datagram Protocol (UDP), Simple Mail Transfer Protocol (SMTP), Post Office Protocol 3 (POP3), Hypertext Transfer Protocol (HTTP), File Transfer Protocol (FTP), and SSH
- Working knowledge of popular cryptography algorithms and protocols such as Advanced Encryption Standard (AES), Rivest, Shamir, and Adleman (RSA), Message-Digest Algorithm (5) (MD5), Secure Hash Algorithm (SHA), Kerberos, Secure Socket Layer/ Transport Layer Security (SSL/TLS), and Diffe Hellman
- Security engineering and architecture work—analysis and engineering of security features of large, distributed systems

- Work with some COTS NIDS/NIPS or HIDS/HIPS tools such as McAfee IntruShield [70] and ePolicy Orchestrator (EPO) [71], or Hewlett-Packard (HP) TippingPoint [72]
- Work with various log aggregation and SIEM tools such as ArcSight [73] or Splunk [74]
- Experience with vulnerability assessment and penetration testing tools such as Metasploit [75] [76], CORE Impact [77], Immunity Canvas [78], or Kali Linux [79]
- For those working in malware reverse engineering, (1) knowledge of assembly code in Intel x86 and possibly other popular architectures, (2) work with malware analysis frameworks such as ThreatTrack ThreatAnalyzer [80] and FireEye AX [81], and (3) work with various utilities that aid in malware analysis, such as SysInternals [82], and tool suites used to decompile and examine malware (not the least of which is IDA Pro [83] [38])
- Experience with programming and scripting languages and text manipulation tools, most notably Perl [84], but also including sed and awk [85], grep [86], Ruby [87], and Python [88]
- For those doing forensics work, knowledge of Windows and other OS internals and popular file systems and work with media forensics and analysis tools such as AccessData FTK [89] or EnCase Forensic [90].

In addition, the candidate screening process should address candidates' soft skills:
- Written and oral communication
- Ability to thrive on high ops tempo, high-stress environments
- Strong team player
- Ability to provide on-the-job training and knowledge sharing to other analysts
- Self-initiative with strong time management
- Solid sense of integrity and identification with the mission.

There is certainly a lot of overlap between CND and general IT and network operations. CND operators must be able to speak the language of general IT. As we discussed, it's the ability to think like the adversary and hunt for anomalous and malicious activity that sets them apart from the general IT crowd. For this reason, it is not reasonable to expect to rotate personnel between CND and non-CND positions. Momentary surges in SOC staffing are usually in response to an incident. Rather than rotating people directly into the SOC itself, it is usually better to leverage other IT personnel in their existing slots, such as with TAs, or turn to partner or superior SOCs for help.

> **Extensive background and skills are shared among CND and general IT operations; however, these staff positions are not interchangeable.**

It is possible and, perhaps, encouraged to rotate staff among various junior positions such as Tier 1 analysis and vulnerability scanning. However, as we ascend in seniority, the amount of ramp-up time increases. For instance, it may be possible to move a Tier 1 analyst to network scanning with just a week of on-the-job training. However, to move a Tier 2 analyst to a Red Team (or back) will usually take longer.

For more details on relevant skill sets for CND personnel, refer to NIST's NICE framework found at [91] and materials from CMU SEI CERT at [8] and [92].

7.1.4 Conclusion

To finalize our discussion, here are some tips for hiring well-qualified analysts:
- Consider the full gamut of qualifications, including both formal education and professional certifications, training, self-study, and on-the-job experience.
- Tailor an interview process that focuses on out-of-the-box thinking through open-ended questions and recognizes fundamental background skills like understanding of TCP/IP protocol, UNIX system administration, and programming.
- Look for personnel who have a mix of related skills like software development, vulnerability assessment/penetration testing, and advanced system administration, who can quickly adapt to CND operations.
- Look for personal traits during the interview that indicate the candidate has a passion for "cyber," strong communication skills, the desire to work with a tight-knit team, an orientation toward the fast ops tempo environment, and a thirst for knowledge.
- Leverage a pay-band structure or contracting model that supports differentiation of tasks and experience levels across all areas of current and planned SOC work.
- Rely on references from existing rock-star resources who have friends interested in working in CND, even if their experience in IT is not security focused.
- When hiring supervisory or management positions, ensure those with management credentials also bring hands-on experience with IT (preferably CND) to the table.
- Utilize a technical qualification or "check-ride" process that each new hire must pass within a certain time period after hire; this ensures all employees can operate with a base level of technical capabilities, depending on their job function

7.2 How Many Analysts Do I Need?

This is one of the most frequently asked questions when shaping the SOC, both from SOC managers and those new to CND. Unfortunately, it is one of the hardest to answer, because there are so many issues at play. In this section, we break down the factors that impact overall SOC staffing, and we leverage the Large SOC models from Section 4.1.2 and Section 4.3.3 in looking at how to staff each part of the SOC.

7.2.1 General Considerations

When consulting available resources on SOC staffing guidance, the most frequent models leverage simple ratios: either number of analysts to number of devices being monitored or number of analysts to number of constituents [93]. The most frequent ratio quoted is one analyst for every 50 to 75 devices [94, p. 9]. Hands-on experience proves it's often not that simple. First, these models work for a given SOC when all other considerations remain fixed. As we have seen in this book, we have to look at the whole picture: people, process, and technology. Second, these models typically have Tier 1 analysts in mind, whereas we wish to address all roles within the SOC: Tier 2, cyber intel analysis, system administration, and so forth.

Let's look at all of the factors that influence SOC staffing levels:
- SOC mission and offered capabilities
- Size, geographic distribution, and heterogeneity of the constituency
- Number of incidents (detected or otherwise) on constituency systems
- SOC organizational model
- Size, coverage, and diversity of SOC monitoring and analytics systems
- Intended and existent SOC staff skill set
- Business/coverage hours offered by each SOC section (8x5, 12x5, 24x7, etc.)
- Level of automation built into SOC monitoring, correlation, and analytics
- SOC funding for staff resources.

Let's recall a few key points. At the start of Section 7, we observed that one skilled analyst with force multiplier technologies such as SIEM can be as effective as 100 rookie analysts with poor tools. In Section 6, we saw that SOCs can have a variety of different capabilities, and, of course, the SOC will attain differing maturity levels for each service it offers. From Section 4.2 and Section 4.3, we know SOCs can come in vastly different shapes and sizes. From Appendix D, we know that an 8x5 position takes one full-time equivalent (FTE), whereas a 24x7 position requires 4.8 FTEs.

As we can see, SOC staffing needs require a more complicated equation than a simple ratio—we have a list of many independent variables, each one having a potentially profound effect on the answer. For new SOCs, few of these factors may be set in stone at initial formation (e.g., when budgets are first cut). For many SOCs, the answer evolves over time, as they grow and mature. SOC managers typically seize one of four different opportunities to grow or shape their staff: (1) in the wake of a major incident that has constituency executives' attention, (2) when the SOC's organizational placement is changing, (3) at the early stages of annual budget planning, or (4) in the wake of a major inspection or assessment.

However, adding people should be done only after exhausting other process and technology options that can act as force multipliers.

> **Throwing more SOC staff at a problem is usually not the best answer. First, consider automating human–intensive processes and seeking more streamlined escalation and response CONOPS.**

If a mediocre analyst's salary is X and a great analyst's salary is X*1.25, but a great analyst gets three times more done than a mediocre analyst, then it's actually more efficient to hire or grow fewer great analysts. That said, finding or growing good analysts can be a major challenge with great demand among a finite pool of qualified applicants, and an operations tempo that sometimes leaves little room for personal growth.

In the following sections, we will examine the primary factors influencing staffing for each section. It is also important to keep in mind that the hours each section may keep can vary. Tier 1 may be required to staff 12x5 or 24x7, whereas the rest of the SOC may maintain a more limited 8x5 schedule.

7.2.2 Tier 1 Analysts

Let's recall our discussion of Tier 1 from Section 2.2 and Section 4.1. We have a team of generally junior folks who perform the lion's share of the SOC's routine tasks: fielding phone calls and monitoring well-tailored event feeds. They have limited time to focus on any one event of interest. Out of any of the SOC's sections, the staffing model here is the most predictable. If we consider the average number of minutes it should take an analyst to evaluate the disposition of an alert and the number of alerts *worthy of their attention* in a given shift, we know how many analysts we need [9, p. 401]. Sort of. Let's recall one of the most important lessons learned when it comes to Tier 1 monitoring:

> **Never ask Tier 1 to monitor a completely unfiltered data feed. This will cause them to spend most of their time clearing benign alerts and becoming numb to what is truly anomalous or malicious.**

With too many events showing up in their dashboards, Tier 1 analysts have two options: (1) furiously acknowledge or skip many alerts without fully analyzing them or (2) hunt and peck for random alerts out of their feeds. Instead, Tier 1 analysts should be presented with discrete views into the data that can be fully evaluated over the course of their shift. The number of alerts the analysts must deal with in a shift is highly dependent upon many factors, most notably the quality and quantity of data feeds flowing into their tools, and the analytics applied to them. And, it is hoped that they are presented with views into the data that are something other than just scrolling alerts. Modern SIEMs provide all sorts of data visualization tools. As a result, preparing any sort of mathematical formula to predict Tier 1 staffing is hazardous at best.

How well tuned is the SIEM? the IDSes? the data feeds? Are all of the alerts unified into one or more SIEM dashboards, or are they split among half a dozen disparate tools? That said, just because we deploy a new sensor technology or add a new dashboard to SIEM doesn't necessarily mean we need to hire more analysts. If we're lucky, we might be able to completely automate certain monitoring use cases and send autogenerated cases directly to Tier 2. Clearly, there is a lot of gray area here.

Also, we need to consider other tasks thrown at Tier 1. In a small SOC, Tier 1 may also do routine cyber intel collection or vulnerability scanning. We may also have staff dedicated to other routine tasks like monitoring constituents' Web-surfing habits or IT compliance-related activities. This certainly adds to their load. The SOC escalation CONOPS will come into play here because Tier 2 may push down handling of routine events such as malware infections to Tier 1, assuming they demonstrate the capability to act upon certain incidents with competence and consistency. On the other hand, we may have to minimize the activities carried out by Tier 1 simply due to physical space limitations—perhaps Tier 1 exists on a cramped ops floor, whereas Tier 2 and the rest of the SOC sits in a back office.

Understanding the range of constituency sizes we have to deal with, Large Centralized SOCs will commonly have between two and six analysts on each shift in Tier 1. This can be quite deceiving to outsiders, because, on a floor tour of a combined NOC/SOC, they may only see the Tier 1 analysts and a floor lead. What they may not notice is the other parts of the SOC residing in back offices, which are just as important to the mission.

7.2.3 Tier 2 Responders

The number of folks needed to fill slots in Tier 2 is most directly related to three factors: (1) the frequency and number of cases passed from Tier 1 or other parts of the SOC (e.g., cyber intel analysis and trending), (2) the amount of time the SOC is able to devote to each case, and (3) the ability of each analyst to turn over cases, which is influenced by their skill and tools.

For instance, some SOCs can be stuck in the response cycle: find intrusion, pull box off network, reimage box. As we discussed earlier in this book, this is tremendously counterproductive. On the other end of the spectrum, we have SOCs that have an advanced adversary engagement, tradecraft analysis, and reverse engineering capability. Furthermore, the SOC may have these capabilities, but they may be split out into a different "advanced capabilities," "threat analysis," or "forensics" section, as described in Section 11.1. Some SOCs will actually host an entire malware catalog and analysis framework, further increasing staffing needs.

Staffing for this section is heavily influenced by the SOC's ability to find (and pay for) staff capable of carrying out Tier 2 analysis, as well as the overall vulnerability and threat profile of the enterprise. If the constituency is getting hacked left and right, clearly there will be a greater demand for incident responders than if things are relatively quiet. In addition, the SOC may feel compelled to spend cycles chasing down IT misuse cases such as users caught surfing porn or gambling sites while at work. It's easy for the SOC to spend resources on these cases because it has the right tools to investigate them, even though such cases should probably be moved to another organization such as security or human resources (HR).

Depending on all of these factors, SOCs may have Tier 1 to Tier 2 seat ratios anywhere from 2:1 to 1:2. This means that for every two Tier 1 analysts on a day shift, there could be between one and four Tier 2 analysts, depending on how operations and escalation are structured. In terms of actual FTEs, this may be more like 5:1 or 3:1, because the Tier 1 floor positions are more likely than Tier 2 to be staffed 24x7.

7.2.4 Cyber Intel Analysis and Trending

This section of the SOC has the most open-ended portion of the SOC mission. Staff is asked to consume as much cyber intel and sensor data as possible, in a never-ending quest to uncover anomalous activity. Therefore, the number of staff needed to support this SOC section is almost entirely dependent upon the SOC's ability to fund and find qualified personnel. Staffing here will also be driven by the SOC's access to cyber intel data. If it has poor data, intel, and news feeds, there won't be much to do. If, on the other hand, the SOC has strong relations with partner SOCs, comprehensive monitoring coverage, and advanced analytics, the opportunities are almost endless.

Overall, this section's staffing will likely maintain rough parity with Tier 2. If either section grows more than twice as large as the other, the SOC CONOPS or staffing plan probably requires a further look. Smaller SOCs will likely have a small trending section, due to limited resourcing. Hybrid tiered, coordination, or national SOCs will likely have a very large trending section because their focus is largely shifted toward watching the adversary instead of watching the network. In the most extreme examples, national-level SOCs may designate a number of sub-teams, each focused on a specific brand of adversary or geographic region.

7.2.5 Vulnerability Scanning

Staffing a vulnerability or network scanning capability within the SOC is fairly straight-forward, in that it is dependent on only a few factors. Some SOCs don't have this capability at all (making the answer quite straightforward). For those that do, we need to consider the number of systems being scanned, the complexity and efficiency of tools that do the scanning and roll up the data, and whether the scanning targets are broken up into two or more disparate networks. Some scanning tools, for instance, work very efficiently in break-ing down scanning tasks and executing them from remotely managed nodes. Others may require a human to manually initiate a scan for each enclave and roll up the results by hand.

SOCs that perform network or vulnerability scanning in house will often have a team of two to five people—possibly more if they have a very large constituency; possibly only one person if their scanning tasks are limited in nature.

7.2.6 Vulnerability Assessment/Penetration Testing

Making staffing choices is in some ways similar to cyber intel analysis and trending. VA/PT activities can be very open-ended for many constituencies. That is, there will always be more work to perform. As a result, the SOC can most likely assign as many people as fund-ing permits, with the following caveats.

First, SOC management is advised not to build up a huge VA/PT section while starving other sections like Tier 2 or trending. Second, the SOC must carefully manage its workload on the basis of the authorities and rules of engagement granted by constituency execu-tives. In environments where the VA/PT team has more freedom of action, it may be able to set the agenda for its operational activities. Third, this team may matrix in personnel from other sections. In order to bulk up teams on large "jobs," this SOC should cross-train staff on defensive and offensive techniques and share knowledge of constituency systems and networks. Caution should be exercised here, as rotating staff out of analyst positions means any cases they were working must either be put on hold or handed off to another staff member. In addition, staff must work on VA/PT engagements with some regularity for their skills to stay current.

This section's capacity can also be directly correlated to staffing. Let's say a SOC calcu-lates that an average assessment requires a team of three, plus one lead, and it takes an average of three weeks to perform a "job," start to finish. That works out to 12 staff weeks per assessment. Four assessments work out to 48 staff weeks of effort, meaning we have a formula that works out to four assessments per year for every FTE, including training and time off. Capacity and staffing projections can thus be made. Granted, it isn't always this simple (some jobs are bigger than others), but at least this is a starting point. The SOC can also use this sort of calculation to predict how often it is able to revisit a given network, site, or program (depending on how its assessments are bounded).

7.2.7 Sensor Tuning and Maintenance

Early in this section, we reviewed the problems using the number of managed devices or sensors as a predictor for SOC staffing. However, there is one portion of the SOC in which such a ratio is definitely usable—the sensor tuning and maintenance section. For sensor maintenance and tuning functions, we can actually see this as a ratio, not of the number of sensors, but of sensor platforms or types. This is due to economies of scale afforded by the central management capabilities found in modern COTS and FOSS CND sensor technologies:

- 0.5 FTE per sensor type of small sensor deployment (< 25 network sensors, < 200 server sensors, or < 2,000 desktop host sensors)
- 1.0 FTE per each moderately sized sensor deployment (25–100 network sensors, 100–500 server sensors, or 2,000–10,000 desktop host sensors)
- 2.0 FTE (or more) per each large-sized sized sensor deployment (> 100 network sensors, > 500 server sensors, or > 10,000 desktop host sensors).

In very small deployments, sensor and signature/heuristic policy management is relatively straightforward. As the number of sensors grows, however, management becomes more challenging, as the variety and number of different deployment scenarios and diversity in rule sets require more management overhead. Adjustments must also be made for sensor platforms that entail greater integration challenges, require constant care and feeding, or need an extraordinary amount of custom rule set creating or tuning. In larger enterprises, it is not unusual to have a team of three or four people devoted just to keeping a HIDS/HIPS suite functioning properly. This is due, in part, to its tight integration with the server and workstation environment. These kinds of labor statistics may, of course, cause some SOCs to reconsider their choices about whether certain monitoring packages are really worth the effort.

This is not the whole story. There are a number of other jobs that the sensor and system administration shop must carry out every day.

With the rising complexity of analytic frameworks, many SOCs also feel compelled to commit staff to maintaining and enhancing these systems. Such jobs could involve maintaining the SOC SIEM or audit collection framework, the PCAP collection and retention systems, or the malware repositories. Part of these jobs will invariably entail specialized platform or database administration work (e.g., data warehouse tuning and optimization). SOCs that make a serious commitment to their SIEM implementation will usually designate one person as a SIEM content manager whose job it is to manage and tune the plethora of custom correlation rules, filters, dashboards, and heuristics built into SIEM. Some SOCs that support a large audit log collection framework may feel compelled to devote one or two staff just to ensuring data feeds don't go down and users' needs are being met. Consider a SOC that is supporting an audit collection architecture that serves many sysadmins and security personnel in a large constituency—this size user base will require dedicated system administration resources.

Depending on how hardened and isolated the SOC enclave is from the rest of the constituency, the SOC will likely need to allocate staffing to these functions, potentially encompassing the following:

- Maintaining SOC analyst workstations, domain controllers, active directory objects, and group policy objects (GPOs)
- Patching SOC infrastructure and sensor systems
- Maintaining internal incident tracking database
- Maintaining SOC network switch, router, and firewall infrastructure
- Updating SOC internal or constituency-facing website
- Maintaining SOC network area storage (NAS) or SAN resources.

Inherent in all of these functions is not only the hands-on O&M of systems but also CM, patching, and upgrades. In fact, some larger SOCs may designate someone separate from the sysadmin lead to preside over document management and configuration tracking.

7.2.8 Engineering

Staffing the engineering section of the SOC is influenced by five factors:

1. The number and complexity of SOC systems in operation
2. How often new capabilities are rotated into operation
3. Whether the SOC has any homegrown or custom capabilities to which it must devote development cycles
4. Where the line is drawn between system administration and engineering functions
5. The amount of bureaucracy the SOC must endure as a result of operating within the engineering and paperwork processes/life cycle of its parent organization.

SOCs that draw a line between system administration and engineering usually have a ratio of 1:1 or 2:1 sysadmins to engineers. In other words, if a SOC has six sensor tuners and sysadmins, a team of three or four engineers may be appropriate. Again, this example is quite arbitrary because many engineering-like functions may be carried out by other parts of the SOC. Such duties include developing new sensor signatures or scripting tasks that were previously done by hand. In many cases, it is very difficult to clearly define the separation between operations and engineering. The SOC is constantly taking on new capabilities in order to maintain parity with the adversary—agility is key.

Many SOCs break down the barrier between system O&M and system engineering entirely, meaning the two teams are fully unified. In such cases, we can simply take the staffing requirements for system administration, inflate them by some multiplier, perhaps 1.5, and calculate the total number of individuals needed for system administration and engineering together. However, this actually masks the efficiencies gained by integrating engineering into operations. A SOC without an integrated engineering function usually receives new or upgraded capabilities with a longer wait and with poorer match to operators' requirements. As a result, operations must devote additional resources to applying bandages and duct tape to problems (i.e., making the tools work as intended).

The point is that by having engineering integrated into ops, the additional staffing requirements to engineer new systems are usually more than made up for by the efficiencies gained, to say nothing of the improvement to the mission. If a SOC pursues this approach, it must be sure to maintain appropriate levels of CM and documentation of its deployed baseline capabilities.

7.3 Minimizing Turnover

Finding and keeping good people is one of the biggest problems SOC leaders face. In an operational environment, turnover is a fact of life. Keeping rock-star analysts is especially difficult because careers in "cyber" tend to pay well and talent is always in short supply.

SOC staff members cite many reasons why they like working in their organization and choose not to move elsewhere. Three of the most common are:

1. They feel like a cohesive, tightly knit team of highly qualified, motivated professionals.
2. They experience new and interesting challenges every day.
3. They believe in the mission of network defense—both its importance and its uniqueness.

In this section, we examine methods for maximizing staff retention, especially among the top performers. We will also touch on methods for coping with the turnover inherent in operations. For more information on these topics, see [95], [96], [97], and [98].

7.3.1 Work Smarter, Not Harder

Some analysts are content to stare at the same stream of log data, day after day, week after week, as long as they feel like they're making a "difference." Our rock-star analysts—the ones we really want to keep—code and script to make their lives easier by doing more through doing less. We can retain both groups by enabling the rock stars to continually push updated methods and procedures to the rest of the workforce.

> **A SOC's only hope to achieve comprehensive, effective monitoring and analysis coverage is to leverage tools that automate the early stages of monitoring, including event collection, parsing, storage, and triage.**

It's easy to fall into the daily grind where every analyst comes in every day and looks at the same data in the same way, without any change. The key is to allocate time in people's schedules for driving improvement to SOC tradecraft, such as automation and analytics. These improvements are then handed to the analysts—typically in Tier 1—who are comfortable with following a daily routine. In larger SOCs, this usually is facilitated by dedicated signature-tuning staff who work with all members of the SOC to identify opportunities for new or better monitoring use cases and implement them across the appropriate tools.

We achieve high levels of automation by maintaining an up-to-date, robust, strongly integrated tool set. Talented analysts—the ones who can think outside the box—expect access to a robust set of tools that match the current threat landscape and give them results in what they consider a reasonable amount of time. For instance, running basic queries against a day of log data should be doable in a matter of seconds or minutes, not hours. Having old and broken tools is a quick way to lose talent.

Here are some techniques that can help the SOC reduce monotony, keeping people interested and excited about the mission and focused on the APT:

- Dedicate SOC staff to two related but distinct goals:
  - Tradecraft improvement such as signature tuning, cyber intel fusion, scripting, and analytics development
  - Performance tuning and content management, ensuring SIEM and other analytics platforms operate in a satisfactory manner for all parts of the SOC.
- Leverage automated prevention capabilities where it is cost efficient and appropriate to do so, thereby minimizing the "ankle-biters" that would otherwise soak up SOC resources:
  - Commonsense deployment of AV and anti-malware at the host and the Internet gateway (discussed later in Section 8.3)
  - Automated content detonation and malware analysis (discussed later in Section 8.2.7)
  - Internet content filtering technologies such as Web proxy gateways.

- Drive improvements to the constituency cybersecurity program as much as possible, so that strategic issues recognized by the SOC are addressed at the right level.
- Make handling of routine incidents repeatable and formulaic as appropriate, so they take up minimal amounts of time and can be handled by more junior staff.
- Refer handling of some routine incident types, such as inappropriate website surfing, to other constituency organizations, as appropriate.

7.3.2 Support Career Progression

A SOC is staffed primarily by people who have and want more technical skills. One of the most desirable traits of an analyst—passion for the job—goes hand in hand with the desire to take on new and different challenges. Those with a solid background in IT but no previous experience in CND can enter a SOC in Tier 1 but may expect to stay in larger SOCs for several years, through a few different career paths, as shown in Figure 16.

Self-motivated people can ascend this ladder, largely through skills they learn on the job. However, a certain amount of formal and informal training is necessary. Let's look at some key opportunities to help SOC members enhance their skill set:

- Informal on-the-job training in tools and techniques, through formal supervisor, coworker, and mentorship relationships, brown bags, and workshops, such as:
  - Focus on a particular actor's TTPs
  - Deep dive on constituency architecture or mission areas
  - Advanced tool use
  - Cross-training on SOC functional areas
  - Interesting cyber news and intel items.
- Formal in-house training programs such as:
  - Initial and periodic "check rides" that ensure staff are qualified for their position
  - Enrichment activities such as computer-based training or slide decks
  - Training scenarios where analysts must pick out activity from a real or synthetic set of log data, using the same tools that the SOC has in operation.
- External training and enrichment such as:
  - Certifications relevant to CND such as SANS GIAC [99] and Offensive Security [100] that cover either defensive or offensive topics
  - Training courses for specialties related to CND:
    - Network forensics and intrusion analysis
    - IDS or SIEM deployment, tuning, and maintenance
    - Media forensics
    - Malware analysis and reverse engineering

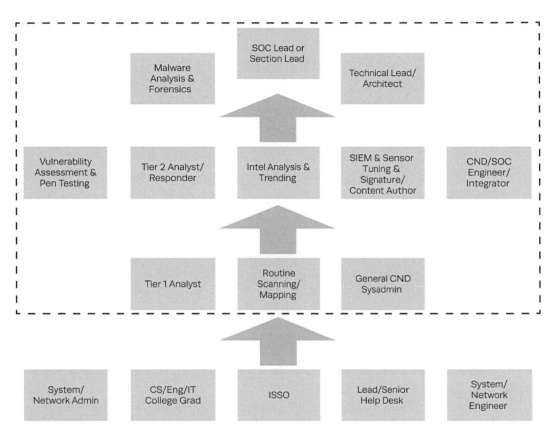

Figure 16. Typical Career Paths Through the SOC

- Vulnerability assessment and penetration testing
- Operating system, database, and network management.
- Professional conferences such as:
  - BlackHat—held at several locations, most notably Las Vegas, Nevada [101]
  - Defcon—held right after BlackHat, Las Vegas, Nevada [102]
  - RSA Conference, San Francisco, California [103]
  - T00rcon, San Diego, California [104]
  - Shmoocon, Washington, DC [105]
  - Skydogcon, Nashville, Tennessee [106]
  - Derbycon, Louisville, Kentucky [107]
  - Security B-Sides, various locations [108]
  - Layer one, Los Angeles, California [109]

- Flocon, Austin, Texas [110]
- PhreakNIC, Nashville, Tennessee [111]
- Hackers On Planet Earth (HOPE), New York, New York [112]
- Hacker Halted, Miami, Florida [113]
- SANSFIRE, Washington, DC [114]
- USENIX, various locations [115].
- Vendor training on specific products in use or being deployed, such as:
  - Product-specific training or certification classes
  - Product-centered conferences for current and prospective customers hosted by vendors such as HP, McAfee, and Cisco.

It is important to recognize that cross-training on job functions greatly enhances employees' ability to extend their job functions, understand other SOC employees' roles, and backfill positions during staffing shortages. In addition, the best network defenders have a keen, hands-on understanding of attack techniques.

For more information on analyst self-development, along with a number of other great tips about hiring and grooming analysts, see [116].

7.3.3 Find the Right Level of Process

We mentioned it before, but it bears repeating—the SOC must find the right amount of structure and freedom in governing the ops tempo and daily routine of its analysts. With too little process, there is no consistency in what the SOC does, how it finds, or how it escalates and responds to incidents. With too much process or bureaucracy, the analysts don't have the time or freedom to pursue the most important leads that "just don't look right" or to rise to the challenge when called upon. If the SOC slips to either end of this spectrum, staff will leave.

What are some good candidates for a formalized process? Here are some examples:
- Overall CONOPS that articulate to constituents, inside and outside, the major inputs and outputs of the SOC escalation process, demonstrating rigor and repeatability in terms of overall cyber incident handling
- Daily, weekly, and monthly routines that must be followed by each section of the SOC:
  - Consoles and feeds that must be looked at
  - Preventive maintenance and health and welfare checks
  - Reports that must be written and pushed to external parties for SA
  - Websites that must be checked for updated cyber intel and news.
- SOPs that describe in detail what is done in response to certain events:
  - Escalation procedures for well-defined, routine, noncritical incident types such as data leakages, viruses, and inappropriate Web surfing

- Escalation procedures for less structured, more unusual, or critical incident types such as root compromise and widespread malware infections
- Downed sensors, data feeds, or systems
- Facility and personnel emergencies (e.g., those initiating a COOP event) such as inclement weather, fire drills, or personnel out on sick leave.

Note that we've left out how the analyst actually evaluates security-relevant data for signs of malicious or anomalous behavior. This is the most critical element of the CND life cycle, and it cannot be turned into a cut-and-dried process. It should be noted that SOC processes (especially SOPs) are more focused on junior members of the staff such as junior sysadmins and Tier 1 console watchers. This is natural, since those newer team members require more structure and routine in their job.

7.3.4 Close the Loop

Let's return to the reasons often cited by analysts for liking their work in a SOC—teamwork and the mission. One of the best ways to encourage this feeling is to provide regular feedback to SOC staff on how their contributions are making a difference.

Probably the most straightforward way to accomplish this is through regular meetings where SOC personnel discuss tactical and operational issues, as well as less frequent quarterly meetings to discuss bigger picture issues. The desire to bring the SOC together must be tempered with mission demands. For instance, with a larger SOC, weekly or daily ops "stand-ups" need involve only SOC section leads.

The second way to accomplish this is to provide feedback (known by some as "hot washes") to the entire SOC on the results of recent incidents. If done properly, this supports several goals:

- Provides evidence that individuals' efforts are having an impact
- Recognizes individuals' accomplishments
- Highlights the importance of each section's contribution to the mission
- Brings to light techniques that can be used across the SOC
- Keeps SOC members informed of nuances regarding incident escalation procedures
- Drives improvement to processes and technologies that are having the most success
- Provides artifacts that can be rolled up into records of accomplishments the SOC can use to justify expanded resourcing and authorities.

The third method is to regularly exercise the SOC. This must be done in a way that minimizes interference with operations while maximizing the exercise integrity and realism. The most successful CND exercises have strong participation by one or two SOC TAs who can formulate realistic attacks, know how to play to the SOC's strengths and weaknesses, and can draw observables and conclusions from their position in ops as the exercise plays out. The more complex and realistic the intrusion—such as a full penetration test against constituency assets or a successful phishing attack—the more opportunities there are for testing the full incident-handling life cycle. Results of SOC exercises should be brought back to management and analysts in a non-hostile, open method, so that everyone can learn from the results and adjust procedures accordingly. Finally, exercise attacks should be carefully coordinated and approved by IT seniors, with a written set of "injects," scripted actions, rules of engagement, and points of contact list.

The fourth way is to maintain regular analyst-to-analyst sharing among SOCs, a topic touched on many times in this book. By regularly sharing "war stories," team members gain a sense of community and belonging. It also helps dispel myths that "the grass is greener" at other SOCs, whereas in reality, most SOCs share the same struggles and find comfort in common challenges and solutions.

7.3.5 Maintain Adequate Staffing

This is probably the most straightforward but challenging prerequisite for retaining a good team. If the SOC is not staffing at the right levels with a qualified cadre of personnel, those who are left are less likely to stay. The reason is obvious—they are overworked and burn out quickly. For more details on this, return to Section 7.2.

7.3.6 Pay Them

Anecdotal evidence suggests that CND is a field dominated by 20- and 30-somethings. Sufficiently talented and self-motivated employees will quickly pick up a variety of highly marketable IT skills in just 12–18 months working in a SOC. For example, it is possible for a fresh college graduate with drive and a CS degree to spend a year or two in a SOC and then jump ship for an annual pay increase of $20,000 or more.

Highly qualified team members are likely to leave for higher paying positions, especially when they don't feel that they have an upward career path in their current organization. Nurturing, mentorship, and self-improvement only go so far. Many SOCs struggle to find and retain the right folks because talent is in short supply and the job market is in the applicants' favor. Add in the fact that many SOCs, especially those in the government, require extensive background checks, further narrowing the field of candidates.

The bottom line is this: a SOC must be able to adequately compensate its employees. This can be especially challenging in government environments or with contracted employment. SOC management should ensure that team members receive adequate compensation and that the SOC is granted different or higher pay bands separate from positions in general IT such as junior sysadmin or help desk. As a result of their higher compensation, SOCs must also be careful to vet all potential hires for strong technical and soft skills.

7.3.7 Have Fun

It is usually not enough to come into work every day, feel that you're having an impact on the mission, and grow professionally. One of the hallmarks of a strong and healthy SOC is that its staff has fun inside or outside of work, on a regular basis. Regular outings for lunch or ordered-in pizza or fast food is a good start. Also, consider team-building and social activities outside work that appeal to the SOC demographic, such as paintball, laser tag, go-kart racing, or local area network (LAN) parties. Keeping the environment inside the workplace easygoing and casual is also key. This means not dressing up in a suit every day and letting some of the analysts listen to music on headphones. Nerf wars are also a good way to let off some steam late on a Friday afternoon.

7.3.8 Cope with It

Turnover is a fact of life for ops centers. Annual attrition rates around 30 percent are not unheard of. As a result, the SOC's ops model must embrace the fact that few team members will stay longer than four or five years. In SOCs that leverage a contractor workforce, the entire staff may be "greened" every three to five years. Here are some tips:

- Keep a living set of SOPs that describe each of the duties, ops tempo, and skills for each work center.
- Constantly stay on the lookout for new hires, leveraging leads from current staff and institutions of higher learning that have a reputation for strong engineering and CS.
- Constantly educate staff on key skills, especially through on-the-job training and, possibly, through archiving presentations and demos for later use.
- Maintain as much institutional and technical knowledge in lead and management positions as possible, compensating for gaps between departures and new hires.

Chapter 8
STRATEGY 6: Maximize the Value of Technology Purchases

···

When many SOCs were first stood up in the 1990s, the number and complexity of tools they had to work with were relatively low, focused largely on network and host-based IDSes. Since then, the marketplace in security products has exploded, as have the adversary's TTPs, resources, and motivation. Today, the SOC must leverage a wide array of capabilities in executing its mission, making the CND mission much more expensive to conduct than in the past. There is no one tool that will "do it all." Each has its limitations, and they often must interoperate in a complex architecture supporting comprehensive monitoring and advanced analytics.

The tools we will discuss to meet these objectives should be familiar to those with even cursory experience in CND. They are:

- **Vulnerability scanners** and **network mapping** systems that help SOCs understand the size, shape, and vulnerability/patch status of the enterprise
- NIDS/NIPS, which are used as tip-offs to user, system, or network behavior that is of concern
- Complements to NIDS/NIPS, including **NetFlow** (which records a summary of network activity), **full-session network capture collection**, and **content detonation devices** (which inspect documents and Web pages for malicious behavior)
- The host counterpart to NIDS, **HIDS/HIPS**, which, in many cases, also include various enhancements and add-on modules such as **AV** and **configuration monitoring**

- A means of gathering, normalizing, correlating, storing, and presenting events from these various sources, such as a **SIEM** system and its less expensive counterpart, **Log Management (LM) Appliances.**

Despite the richness of features and selection found in these tool suites, their cost, integration, and use continue to be a pain point for virtually every SOC.

> **Cost-effective technologies needed to mount a competent defense of the enterprise are widely available; issues of people and process are usually what hold SOCs back from using them effectively.**

In our sixth strategy, our goal is to extract the maximum value from each technology purchase we make, *with respect to the adversary and the SOC mission.* In order to fulfill this, we should:

- Maintain cognizance over the entire threat landscape and what is most relevant to the constituency—what will be the most damaging, and which are most likely to occur.
- Consider the overall value of each tool in terms of visibility, cyber attack life cycle coverage, and longevity.
- Focus on a discrete set of tools that provide maximum value and avoid overlap in functionality where redundancy is not needed.
- Pursue a rigorous requirements-driven approach based on operator feedback.
- Carefully manage expectations of IT executives regarding the virtues and limitations of tools under consideration—there is no panacea.
- Ensure the SOC has the expertise to exploit the full capabilities of the tools chosen.
- Practice continual improvement over the lifetime of each tool by dedicating resources to tuning and analytics and building custom use cases and content.
- Ensure that the tools chosen fit into a carefully designed monitoring, analysis, and response architecture.

8.1 Understanding the Constituency

In order to defend an enterprise, we must first understand it. Monitoring tools such as network and host sensors get the lion's share of SOC analysts' attention. However, the prerequisite for using these tools is having some "local context" for the hosts and networks they monitor.

Tools used to gain this understanding are, in many cases, owned and operated by organizations other than the SOC. As a result, many SOCs overlook them, even though a substantial portion of the SOC's SA can be drawn from data contained or produced by them. Moreover, because they have already been stood up, they can be leveraged by the SOC with little or no cost.

In this section, we will discuss tools and techniques that help the SOC meet the following three objectives:

1. Understand what hosts are in the enterprise, what is running on them, and how vulnerable they are to attack.
2. Understand the network topology and connection between hosts and between enclaves.
3. Draw key connections between IT assets and their supported mission functions.

8.1.1 Asset Information

Enterprises of all sizes must keep track of their property inventory—what it is, where it is, who owns it, when it was purchased, and so forth. In addition, one hopes that a centralized means to roll out software updates and patches, such as Microsoft System Center Configuration Manager [117], is available. Some tools, such as Radia [118] or Symantec Management Agent (also known as Altiris) [119], will actually reach out to end hosts and query system settings and files resident on the hard drive to determine the ground-truth patch status of a system.

SOCs that perform direct monitoring of constituency hosts and networks should have read-only access to the data contained in these asset tracking and management systems (through direct console access, database extracts, regular reporting, or a combination of all three).

Ideally, these systems will provide a wealth of current asset data to the analyst, including:

- Host name
- Media access control (MAC) address
- IP address
- OS and version
- Service pack and patch level
- Installed and running software
- Hardware details and configuration
- System settings
- Purchase date
- Personal owner, if applicable
- Organizational or project association, in some cases.

Consider the common situation where an analyst is looking at an attack that has hit several hundred systems across the constituency. All the analyst may know are the IPs of the victim hosts. Where are these systems physically located? What do they have running on them? Are they possibly vulnerable based on their service pack level? Data from a robust asset management database can answer these questions.

Vulnerability scanners (which we discuss later in this section) can also provide similar data; however, chances are IT operations already has a robust asset management and tracking database in place. While the SOC could deploy its own tool to collect this data, it could leverage an existing tool for free. Many SOCs will take extracts of asset data and bring them into their monitoring systems, or they will build workflow tools that allow an analyst to query an asset management system on the basis of the details of some event data. SIEM, for instance, has robust support for these approaches.

8.1.2 Network Mapping

To understand the topology of the IT enterprise, the SOC typically turns to network maps. Different network maps can depict the enterprise at varying levels of abstraction. Some are focused on small subnets, site networks, and enclaves. Typically, such maps will provide details on edge assets such as access switches and servers—with the exception of Internet perimeters and DMZs, which are usually needed only in response to a given incident. Other network maps depict the enterprise at a much higher level, showing WAN topology, major routers, and perimeter points of presence (PoPs) but leaving out the end host. These are more frequently needed for analysts to comprehend what they are monitoring. SOCs are not usually the custodian of network maps. Therefore, keeping an up-to-date collection may be a struggle, and the SOC must lean on personal connections with system owners and network engineers.

Network maps are normally generated automatically by a network monitoring program or network scanning tool or are manually drawn with a computer diagramming program such as Microsoft Visio™ [120].

Drawing a network map by hand is a fairly unglamorous, tedious task carried out by knowledgeable network and sysadmins who recognize the importance of having accurate depictions of the assets they manage. If analysts can gain access to current, accurate network maps, they have already made a lot of progress in understanding the constituency, at zero cost to the SOC. In fact, some SOCs can actually play an active role in pooling collections of network maps or by providing updates to consolidated high-level diagrams.

To augment manually rendered network maps, many enterprises will implement automated network scanning systems such as Lumeta IPSonar [121]. While each tool has its own virtues and drawbacks, they all have some key architectural commonalities. As we see in Figure 17, scanner nodes are placed in key locations throughout the network. Each node will look for networked devices and their configuration data within a user-defined IP range.

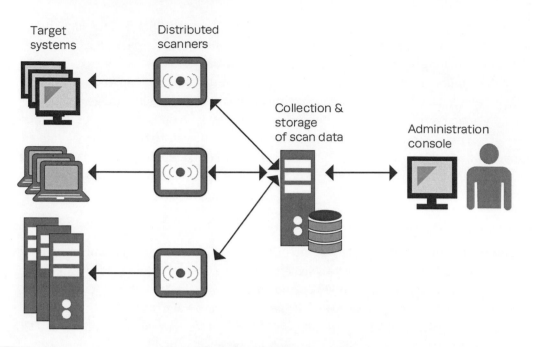

Figure 17. Basic Network Discovery Architecture

This data is then brought back to a central collector where the user can manage the system and interact with the collected data. By analyzing properties of scan results, such as trace routes [122], the system can infer the network topology between it and the scanned assets.

Maps generated using either manual or automated techniques can embed detailed configuration data about depicted assets or include links to asset data resident in an asset tracking or network management database located elsewhere. This is principally done by either remotely querying end devices for key configuration data through Simple Network Management Protocol (SNMP) or placing a management software package on the end host. The latter approach provides a richer set of data and more robust management but at the added cost of deploying and maintaining a piece of software.

More detailed techniques for capturing network data are outside the scope of this book. However, it should be noted that the author of a network map always has to balance completeness with size, complexity, and readability. Moreover, for many network map authors, network maps are a labor of love completed during spare time. As a result, their maps can easily fall out-of-date due to the rapid pace of change for most enterprise systems. Keeping maps updated and correct is a regular job.

We summarize strengths and weaknesses of network mapping techniques in Table 8.

Table 8. Automated Versus Manual Network Mapping

Quality	Automated	Manual
Cost to implement	Moderate, given the size of the constituency; enterprise tools can get pricey, and deployment incurs certain costs.	Minimal to none; little beyond the cost of a few Microsoft Visio (or similar tool) licenses
Cost to execute and sustain	Small to moderate, depending on frequency of scans and complexity of tool used	Small to moderate if done regularly; building, consolidating network maps is seen as an inherent function in network CM; resources must be consciously devoted to it. Bringing antiquated network maps up to date is more costly because fundamental aspects of network topology and configuration must be rediscovered by hand.
What it captures	Networked hosts and interconnectivity, some router/switch configuration—regardless of whether hosts are documented and baselined	Known network devices (switches, routers, firewalls, servers), connections, perimeters, networks—anything the drawing authors felt necessary to include, even if they crosscut levels of abstraction
What it misses	Any significant level of context mapping to the mission; systems not visible from the vantage point of the scanners, due to firewalls, Network Address Translation (NAT), VPNs, network virtualization, or tunneling; configuration data for assets to which it has no access or privileges.	Anything the authors did not know about, such as undocumented assets or changes to the network; in absence of additional automation or scanning plugins, detailed configuration data for large numbers of assets because it would take the map author a very long time to capture said data.
Aging	Deployment and use of scanners must maintain parity with differing networks and enclaves within constituency, otherwise results are only as old as the latest scan.	Network maps are out of date as soon as they are drawn; vigilance drives their correctness and completeness; a diagram may be marked as recent, but portions of the data captured may actually be quite dated; full scrubs of network diagrams can be laborious.

Table 8. Automated Versus Manual Network Mapping

Quality	Automated	Manual
Scalability considerations	Can capture vast quantities of asset data across large networks not practical through manual means; diagram complexity can become a problem if not captured through a hierarchy of drill-down maps.	Drawing each map takes time, so maintainers are forced to include only what is important; depicting thousands of individual nodes on one map is usually impractical.
Impact to systems and networks	Operators must architect and execute scans in a way that doesn't overly obligate resources; used incorrectly, a self-inflicted denial of service (DoS) is very possible.	Little to none. Systems are not necessarily touched.

It should also be noted that there is a third, less common method used to draw network maps. Earlier, we mentioned that manually or automatically generated maps could hot link to asset configuration data directly from symbols on the map. It's also possible to generate a map entirely by analyzing the configurations of various assets such as routers, switches, and firewalls. At least one vendor, RedSeal, offers a network-mapping product that leverages this strategy [123]. This approach has many of the same virtues as the automatic scanning methods we discussed above—it reflects what is actually deployed and running. However, one of its drawbacks is that the network map has blind spots where we do not have rights to interrogate devices for their configuration—a problem that conventional automated mapping does not necessarily suffer from.

Ultimately, each approach to network mapping can be seen as complementary. That said, anecdotal evidence suggests that manually drawn network maps are favored very heavily by both network and security personnel because good hand-drawn network maps capture what is of interest to the reader. Machine-drawn network maps, on the other hand, don't usually emphasize significance, mission, or logical grouping of assets in the same way. In either event, the centralized and distributed SOCs must maintain a strong relationship with network maintainers so they can keep tabs on the latest structure of the network.

8.1.3 Vulnerability Scanners

Consider a situation where an exploit has been released against a popular piece of software. This could pose a serious problem to the constituency if it runs that piece of software without the latest patches, or it may not be an issue at all if the exploit is old and patches are up to date across all systems. Asset management systems may track patch status of systems, but, in many cases, they don't provide robust or comprehensive measurement of whether patches were actually applied. Moreover, not all systems may be reached by an asset or patch management system. In order to address this challenge, the SOC can turn to an enterprise-class vulnerability scanning platform to provide ground truth answers.

From an architectural perspective, vulnerability scanners function a lot like the network scanning tools described above. A user interacts with the system through a central console commanding one or more distributed scanners. Each scanner node is deployed within logical or physical proximity to targeted nodes. Using Windows domain, Lightweight Directory Access Protocol (LDAP), or local system credentials, they interrogate networked systems for system configuration details that provide evidence of their patch status and other security-relevant configuration data. These results are collected by the management server, recorded in a database, and provided to the user on demand. The system provides various options to the user about the depth and breadth of the scan, allowing, for instance, a scan to be run only for certain vulnerabilities.

Vulnerability scanner systems stand apart from network mapping systems because they interrogate the end host for configuration data and provide little insight on the topology of the network. They also gather some data that a host-based sensor can gather. However, vulnerability scanners do not rely on a software presence on the end host, *by design.* There are several tools in the vulnerability scanning market space, such as Rapid7 Nexpose [124], Tenable Nessus [125], and eEye Retina Network Scanner [126].

Vulnerability scanners can also be configured to execute shallower, quicker host "discovery" or simple open-port scans. These don't produce complete results on running applications or vulnerable applications as a full scan would, but they can execute in far less time. If a SOC is lacking a full-fledged vulnerability scanner tool, it may turn to a port scanning tool [127], the most popular of which is Nmap [128], [129].

As we discussed in Section 2.4, a SOC will often perform vulnerability scanning on a portion of constituency networks. By doing this, the SOC functions as the messenger, providing independent verification that IT operations is patching systems within required timelines. While the SOC can be seen as the "bad guy" by IT ops, it is balanced out by the fact that the SOC has a ground-truth understanding of the constituency's vulnerability status, and it is the "go-to" organization that provides this SA to various stakeholders.

Vulnerability scanners are known to produce false positives and false negatives, though they are not known for the same false-positive rates as most IDSes. Like asset data, their results can be brought into SIEM or a full-fledged asset management system for correlation. Understanding their false-positive rates and types is important when correlating against other data types, such as in SIEM.

Finally, it is important that a SOC receive the appropriate approvals (e.g., from the CIO) and provide notification prior to conducting vulnerability, network, or port scan activities. These scans can be very disruptive to legacy systems, which may not respond in a deterministic manner to the nonstandard network traffic sometimes emitted, especially when performing OS and application fingerprint scans. Care should be taken to exclude such systems when necessary. SOCs should have a standing agreement with IT executives and operations leads as to the nature, frequency, and source of scanning activity. Particularly disruptive scans should be preceded by a "heads-up" email to relevant network admins and sysadmins.

8.1.4 Passive Fingerprinting

Considering that vulnerability and port scanning can cause nontrivial disruption of system and network services, the SOC may seek out other approaches to understand what's running on the network *in real time*.

Open source utilities such as P0f [130], XProbe2 [131], and at least one popular commercial tool, Sourcefire Real-time Network Awareness [132], leverage a passive approach to identifying what's on the network. By placing them next to, or on, network sensors that the SOC has deployed throughout the enterprise, the SOC can gain this added visibility at little added cost. They will often leverage a library for packet capture (libpcap) to pull packets off the wire, thereby offering a familiar operating environment. In fact, some commercial NIDSes can be configured to produce passive OS and application fingerprinting results in the form of alerts, just like a signature match.

There are a few caveats to keep in mind. First, tools like ettercap [133] can provide passive fingerprinting but have several other uses that are more oriented toward penetration testing activities. Some AV packages may flag them as "hack tools," whereas the SOC has legitimate uses for them. Second, passive tools will, of course, not see running hosts or services that do not actively communicate over the link being monitored. As a result, such a tool may produce results that are mostly *correct*, but they are certainly not *complete*.

For more information on passive fingerprinting, see [134].

8.1.5 Mission and Users

So far we have covered a variety of technologies and techniques the SOC can leverage to understand the technical attributes of the constituency. Recalling our discussion of the fundamental components of SA from Section 2.5, SA is, at best, incomplete without drawing both actors (users) and the mission context into the picture. Understanding network topology and host configuration is a more straightforward task by comparison.

Understanding their constituency's mission, lines of business, and organizational structure is a challenge for many SOC analysts due, in large part, to the demands of their ops tempo. This works to their disadvantage because the significance of potential incidents can only be evaluated when the mission and people aspects come into focus. While many tools such as SIEM advertise the ability to integrate mission and business context into the tools, this information almost always must be captured and entered into the system manually. There is no tool that can scan the network and automatically say, with consistently high confidence, that "This system is a development box" or "This network belongs to accounting" without a human first defining such relationships. Furthermore, tools designed to capture dependencies between mission and IT assets are largely in their infancy.

Here are some techniques we can leverage to address these challenges:

- Require new SOC personnel to attend their constituency's mission introductory course; if one does not exist, consider integrating this into the SOC training program.
- Tier 2 analysts may accumulate knowledge of key connections between IT and mission over time; consider regular knowledge sharing among team members.
- Network maps, especially those drawn by system owners and project engineers, will often include context regarding systems mission role; asking a few pointed questions of a network admin or engineer based on careful examination of a network map can be very helpful.
- HR databases and identity management systems often include annotations regarding each user's organizational alignment and business function. High-end SIEM systems and insider threat-monitoring tools can actually gather this data and perform advanced correlation on alerts in the context of user roles.[1]

---

1 As with collecting any sort of data that has personally identifiable information in it, the SOC should be careful to respect applicable privacy laws when gathering and retaining records from Human Resources (HR) systems and directories.

- Some constituencies will actually capture information about each of their departments and subdivisions through a structured website or database, which can be browsed by employees or queried by other systems.

8.2 Network Monitoring

For most SOCs, the traditional cornerstone of their incident detection and data collection framework is a fleet of network-based sensors deployed across the constituency. An analyst needs three things in order to perform competent network monitoring:

1. An initial tip-off capability such as a signature- or behavior-based IDS. This includes the ability to leverage custom signatures and full details on the signature or behavior that fired (e.g., signature syntax).
2. NetFlow records that show a summary of communications to and from the hosts listed in tip-off information, days or weeks before and after the tip-off fired
3. The packet capture for the packet(s) that triggered the alert, preferably for the full session, in the form of libpcap-formatted data (PCAP).

With all three of these elements, along with effective analytics and workflow, the analyst can identify anomalous or malicious activity and determine whether further action is warranted. Ideally, both the NetFlow events and IDS alerts should be indexed against the PCAP data, allowing seamless workflow for the analyst. Few products do all three of these well. This compels the SOC to combine a number of different products in its architecture. Best practices for most modern SOCs mean they will augment these three passive systems with in-line preventative capabilities, such as a NIPS or content detonation device.

In this section, we cover each of these systems and show how they work together in one coherent architecture.

8.2.1 Intrusion Detection Systems Overview

Intrusion detection systems, as stated in [42], are:

Hardware or software products that gather and analyze information from various areas within a computer or a network to identify possible security breaches, which include both intrusions (attacks from outside the organizations) and misuse (attacks from within the organizations).

Adapting [42] further, **network IDSes** are IDSes that capture and analyze network traffic for potential intrusions and other malicious or anomalous behavior.

IDSes have been around for a long time and are discussed at length in various materials such as [2], [46], Appendix B of [9], [135], and [5]. Figure 6 on page 36 shows the classic function of a NIDS. The IDS evaluates network traffic in real time against a signature policy, definition of acceptable/normal behavior, or some other set of heuristics, generating alerts that are sent to a user console and data store. We will defer an in-depth discussion of NIDS to the above sources, and focus instead on the practical architectural considerations for the SOC.

Every time an IDS detects activity that is of concern, such as a match against one of its signatures, it will generate an alert. This alert should contain details necessary for the SOC analyst to understand what the alert means and what to do about it. Most typically, an IDS alert (specifically one from a signature-based network IDS) is composed of the following fields:

- Event identification number (ID)
- Date and time (sometimes down to the millisecond) that the signature fired
- Source and destination IP
- Source and destination UDP or TCP port
- Signature name and/or signature ID
- Event severity
- A textual description of the signature or a link to an external repository or database with details on the signature, such as Common Vulnerabilities and Exposures (CVE) entry and signature description
- Bytes sent and bytes received for the total network session that the signature fired on
- Additional contextual information, possibly including protocol-specific fields such as with SNMP, SMTP, POP3, HTTP, FTP, SSL, or Common Internet File System/Server Message Block (CIFS/SMB).

IDS alerts sometimes also include a reference to the raw PCAP for either the packet(s) that triggered the signature or the entire session. Rather than delivering all of this data to the analyst along with every alert, a reference may be included in the event data such that the analyst can retrieve the PCAP on demand.

NIDSes come in both software and hardware (appliance) form. They can leverage signature or anomaly detection methods that we discussed above or, in some cases, a combination of both. NIDSes can also sit in-line between the attacker and target, not just alerting on malicious activity but actively blocking it; these systems are NIPSes.

In a large enterprise, a SOC will typically have multiple NIDS or NIPS sensors deployed at major choke points such as network perimeters, Internet connections, and, in some cases, at major core switches and routers. The NIDSes respond to tasks from a central manager, such as signature updates, and also send the alerts generated by their detection engines to the manager. An analyst can log in to the central manager, usually through a Web client, and view alerts and system status and manage sensor policies. This architecture is shown in Figure 18.

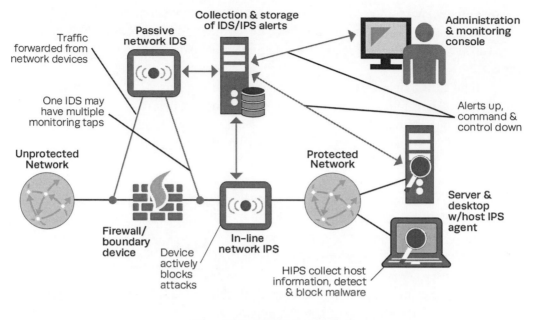

Figure 18. Typical IDS Architecture

So far, we've discussed general-purpose IDSes that must balance attention to all protocols, sometimes at the expense of deep inspection of a specific type of traffic. These systems comprise the vast majority of network sensing and protection capabilities deployed by mature SOCs but don't fill in all the gaps. As a result, there are protocol-specific detection and prevention capabilities we can bring to bear, which sometimes blur the line between IDS/IPS and firewalls. These products focus on one specific protocol such as Extensible Markup Language (XML), relational database management system (RDBMS) SQL traffic, Web traffic, or Web services traffic. In these cases, vendors will typically build robust protocol reconstruction and decoding engines in order to detect and, potentially, block malicious activity that would slip by general-purpose IDSes. Such devices are most appropriate for instrumenting critical services exposed to large user populations.

Each of the different types of IDSes has its strengths and weaknesses. These are summarized in the Table 9.

There are several attributes of a good IDS, as described in [2, pp. 256-258]. To today's network defender, perhaps the most important function of an IDS is to detect attacks that the enterprise is not comprehensively protected against. This occurs from the time an attack is discovered to the time when systems are patched, or other mitigations are put in

Table 9. Advantages and Disadvantages of Intrusion Detection Elements

Characteristic	Type	Advantage	Disadvantage
Detection Method	**Behavior-based:** Known as anomaly detection	Behavior-based IDSes can detect previously unknown attacks and misuse within a session, prior to a specific attack being publicly known (e.g., with "zero days").	They are complex and prone to false positives. They require longer ramp-up times for IDSes to learn baseline system behavior. Networks or systems with frequent changes and activity surges may be difficult to profile for effective monitoring.
	Knowledge-based: Known as misuse or signature-based detection	Signature-based detection is fast and sometimes has a lower false-positive rate than behavior-based detection. Signature-based IDSes can detect known attacks immediately.	Signature-based IDSes can only detect known attacks. If signatures are not updated, new types of attacks will most likely be missed, putting attackers and defenders in a game of "cat and mouse." They are prone to false positives. They are blinded by content obfuscation or protocol encryption.
Source	**Network:** Detect activity from network traffic at perimeter or core monitoring points	A NIDS can monitor a large range of systems for each deployed sensor. NIDSes should be invisible to users.	A NIDS can miss traffic and is prone to being spoofed, attacked, and bypassed. A NIDS often cannot determine the success or failure of an attack. NIDSes cannot examine encrypted traffic.
	Host: HIDSes monitor OS and interactive user activities. Sensors are often software agents deployed onto production systems.	A HIDS will not miss an attack traffic directed at a system due to missing, encrypted, or obfuscated network traffic, assuming the HIDS is capable of detecting it from system activity or logs. A HIDS can help determine the success or failure of an attack. A HIDS can help identify misuse by a legitimate user. A HIDS often bundles other capabilities such as host integrity/assurance monitoring.	HIDS software could be disabled or circumvented by a skilled attacker. Tuning a HIDS can be challenging since many have easily circumvented detection mechanisms or they require nontrivial training (and re-training) on normal system behavior. A HIDS often requires privileged access to the system in order to prevent or block misuse. Incorrectly configured HIDSes can easily interfere with correct host operation.

Table 9. Advantages and Disadvantages of Intrusion Detection Elements

Characteristic	Type	Advantage	Disadvantage
Response Mode	**Active:** Active IDSes, called IPSes, react by terminating services or blocking detected hostile activities.	IPSes are well matched with signature-based IDSes because of the need for well-known attack definitions. IPSes can prevent or reduce damage by a quick response to a threat or attack. No immediate operator intervention is required.	IPSes require some control of services being monitored. IPSes require careful tuning in order not to block or slow legitimate traffic or host activity.
	Passive: Passive IDSes react by sending alerts or alarms. These do not perform corrective actions.	Easier to deploy False positives do not negatively impact constituency.	Requires operator intervention for all alerts. This adds time to interpret, determine corrective action, and respond, which could allow more damage to occur.

place. This reinforces the value of anomaly-based IDSes that don't depend on signatures and the importance of keeping signature-based IDSes up-to-date.

As we can see in Figure 19, an IDS is most valuable between the time an exploit is put in use by the adversary and when the exploit is patched against. We can also see the gap inherent in signature-based IDSes where signature implementation lags behind when an exploit is in use. Because a signature-based IDS is usually more precise in spotting an attack, once the corresponding signature is deployed, a signature-based IDS may be regarded as more valuable than a heuristics-based one with respect to the exploit in question. As use of the exploit wanes, the value of that IDS with respect to the particular vulnerability also diminishes.

For more information on placement of IDS/IPS technologies, see <u>Section 9.1</u> and <u>[136]</u>. For detailed comparisons between different IDS/IPS products, see <u>[41]</u>.

8.2.2 Implications of Prevention

So far, we have focused on passive IDSes. While these devices provide awareness and tipping to the SOC, they don't actually *stop* anything from happening—they just produce data for ingest by other tools and analysts. It would be a lot better if we could deploy a technology to actually block attacks in real time. NIPSes provide this capability but require even

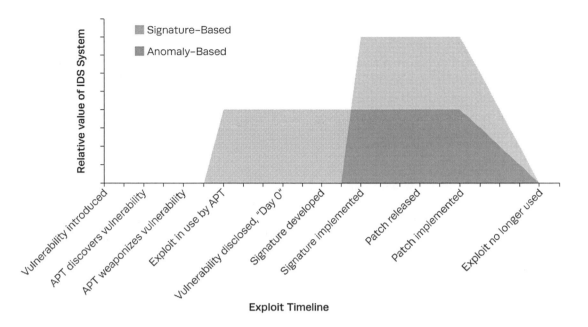

Figure 19. IDS Signature Age Versus Usefulness in Detection

greater vigilance and upkeep than NIDSes because of their potential impacts to network performance and availability. Before we move on to other passive monitoring techniques, we will drill down on "response modes," from above, and analyze what it means to go in-line.

In order to effectively leverage a NIPS's value proposition, the SOC must respect some of the operational realties of operating them. These include:

- **False positives.** Given a SOC's experience with the high false-positive rate associated with IDS technology, NIPS administrators are justifiably cautious. Consider that each false alarm results in blocked traffic. If not careful, the NIPS administrator can inflict a very serious DoS. As a result, many SOCs will be very careful about which NIPS signatures they turn to block, doing so only after several days or weeks of use in alert-only mode. This places a great deal of emphasis on choosing a NIPS with a robust protocol analysis and signature detection engine.

- **Response choices.** Different IPS technologies offer different means to respond to malicious traffic. One common method is to send a TCP reset to both hosts involved in a network connection, which, unfortunately, is not a good idea. Attackers, being malicious, probably expect this behavior and will take advantage of it in two ways: (1) they can simply ignore the reset and continue the conversation and (2) they now know there is probably an IPS in front of their victim. Other IPSes will implement blocks by automatically updating a firewall or router's access control list (ACL), blocking the network communication in question. This is problematic because they can make a mess of router and firewall configurations. The best IPSes implement the block by themselves—simply dropping the offending packet and every subsequent packet between the attacker and victim host(s).

- **Response actions.** Let's consider a situation where we have an IPS that properly blocks packets in an offending network connection and takes no other actions such as TCP resets. How long should we ban the attacker from communicating or with whom—just the victim or anyone on the network? Architectural aspects of the network, such as NATing, can make this more difficult. Imagine an IPS that sees attacks as originating from a Web proxy or NATing firewall. The attacker may appear to the IPS as the firewall, whereas, in reality, it is a host somewhere on the other side. But, because we're blocking traffic, we're now dropping packets from our own firewall, thereby inflicting a DoS. Careful placement of the IPS in order to affect correct response action is critical.

- **Presence.** A NIPS should not advertise its presence to systems on either side of the network connection. This means not sending out traffic to the attacker and target, and not even having a MAC address. In other words, the device should simply appear as a "bump on the wire." That said, even the best IPS may disclose its presence simply by doing its job. Skilled attackers can detect an in-line NIPS by using very old attack methods against a target network. Ordinarily, these attacks will almost certainly fail because the targets are well patched. However, the NIPS will do its job by blocking the attacker from any further communication to the targets. As a result, the attacker now knows a NIPS is present and can change attack techniques. Therefore, the SOC may choose response actions that only block specific attacks rather than banning attacker IPs completely.

- **Latency and bandwidth.** Being in the middle of network traffic, a NIPS may have an undesirable impact on network performance. A poorly implemented or undersized NIPS can introduce latency into network communications or inadvertently throttle bandwidth or traffic that passes by. The SOC must be careful which NIPS products it chooses for a given network connection in order to avoid this problem, especially for high-bandwidth links.

- **Cost of decoding.** A NIPS may operate at its advertised speeds only with a synthetic set of protocols or with certain decoder modules turned off. For instance, a NIPS advertised as 10 gigabit (Gb)-capable may operate at two Gb with its HTTP decoding module turned on, because reconstruction of the HTTP and the logic needed to detect attacks in the protocol are very expensive [137]. Or a NIPS may advertise compatibility with an open signature format (such as Snort), but use of this capability will slow the NIPS's decoding engine significantly. Moreover, a lot of NIPSes offer "packet shaping," "traffic shaping," or firewall-like capabilities. Use of these features will likely introduce latency into network traffic. The SOC must carefully assess whether this latency has a meaningful impact on network services.

- **Content modification.** Some of the fanciest IPS products, especially those advertising information leakage and industrial espionage prevention features, may not only block traffic but *actually modify it*. This could include redacting material from within Web pages or Word documents *as they fly across the wire*. This is a complex undertaking because it requires doing protocol reconstruction and modification at an even higher level of abstraction than network traffic. For instance, consider that an application expects a certain number of bytes, and that this expectation is embedded as a field in the protocol (as with a checksum). If the transmitted content is modified, the checksum must also be changed. Can the NIPS do this without incurring a significant delay in network traffic? These techniques can be problematic and are recommended only for the most well-resourced SOCs.

- **Single point of failure.** If the NIPS is in-line, what happens if it breaks or loses power? Good IPSes will have features (or accessories) that allow them to fail open (i.e., even if it malfunctions or loses power, traffic will continue to flow). This may sound like a bad idea, but remember that an IPS is usually not the only device protecting a network from the outside world. There should be routers and firewalls nearby that are set to a default deny policy. Most commercial NIPS vendors build in fail-open features, whereas extra precautions or architectural changes must be considered when implementing an open source IPS on commodity hardware.

- **Involvement in network operations.** When the SOC deploys any sort of in-line capability anywhere in the enterprise (NIPS, HIPS, or otherwise), it becomes a *de facto* player in network operations. It is very common for the SOC's equipment to be blamed for any sort of problem that crops up (e.g., network outages or slow performance), even if the relationship between the SOC equipment and the actual problems is far-fetched. A SOC will sometimes find its equipment has simply been disconnected because network operations asserted that the SOC's gear caused some problem. A SOC must be vigilant in watching its device status and data feeds to catch issues, and it must constantly work with network operations to ensure its equipment is well behaved.

- **Price.** A NIPS cannot tolerate performance or availability problems that a passive IDS can. NIPS vendors are, therefore, compelled to build robust, sometimes custom hardware platforms into their products. Whereas a commodity NIDS sensor may cost less than $10,000, a NIPS operating at the same speeds may cost substantially more. Most NIDSes sold today are dual purpose; they are able to operate both passively off to the side or actively in-line. That said, running in-line uses twice as many ports on the device, thus doubling the effective cost of running in-line for a given set of network links.

Despite these challenges, some SOCs have found NIPSes to be useful tools. The gap between exploit release and patch deployment presents a period of serious risk to the enterprise; sometimes this is measured in hours, other times it is measured in weeks or months. A few well-placed IPSes may provide protection during periods when constituency systems would otherwise be vulnerable.

Unfortunately, many SOCs never get their NIPS into an in-line, blocking mode. On the basis of discussions with various SOCs and IPS vendors, some, if not most, NIPS devices remain in passive span/tap mode for most of their operational lives. There are several potential causes for this: (1) the SOC doesn't have enough organizational authority and operational agility, (2) the SOC isn't confident enough in its signature tuning, or (3) the SOC simply decides that in-line blocking mode isn't worth the perceived risk of a self-inflicted DoS. There are, of course, other methods to block attacks in-line that some SOCs favor instead of NIPS; content detonation devices and host IPS, discussed in Section 8.2.7 and Section 8.3.3, respectively.

8.2.3 NetFlow

Whereas an IDS looks at the entire contents of network traffic, we need a capability that summarizes all network traffic, with little performance overhead. Among the complements to IDS alerts are **NetFlow records** (often referred to as **flow records** or **flows**). Rather than recording or analyzing the entire contents of a conversation, each flow record contains a high-level summary of each network connection. While the development of the NetFlow standard can be attributed to Cisco, it is now used in a variety of different networking hardware and software products [138] and is an Internet Engineering Task Force standard [139].

One can think of a NetFlow generation device as like a telephone pen register. A pen register is a device that produces a listing of all phone calls that go in and out of a particular phone line, including the time, duration, caller's phone number, and recipient's phone number [140]. However, a pen register is just a summary of what calls were made. It doesn't include the contents of the call—what was said. NetFlow is like a pen register for TCP/IP network traffic. Simply put, NetFlow is to PCAP collection as a pen register is to a transcript. NetFlow doesn't say *what* was said, it simply indicates that a conversation took place.

While different NetFlow generation and manipulation tools are available, generally speaking, each flow record provides the following information:

- Start time
- End time (or duration since start time)
- Source and destination IP
- Source and destination port
- Layer 4 protocol—TCP, UDP, or ICMP
- Bytes sent and bytes received
- TCP flags set (if it's a TCP stream).

Whereas the contents of a network connection could be gigabytes (GB) in size, a single flow record is less than a few kilobytes (KB). The power of NetFlow, therefore, is found in its simplicity. NetFlow record collection and analysis is regarded as an efficient way to understand what is going on across networks of all shapes and sizes. It is critical to understand that NetFlow records do not generally contain the content of network traffic above OSI layer 4. This is a blessing, because (1) little processing power is necessary to generate them, (2) they occupy little space when stored or transmitted, (3) they are agnostic to most forms of encryption such as SSL/TLS, and (4) just a few flow records can summarize GBs or perhaps terabytes (TBs) worth of network traffic. On the other hand, this is a curse, because the flows capture nothing about the payload of that traffic. Whereas an IDS consumes significant processing power to alert on only suspect traffic, NetFlow generation tools consume little processing power to summarize all traffic.

Interesting clues can be generated from NetFlow alone—for example, "Hey, why do I see email traffic coming from a Web server?" or "One workstation was seen transferring vastly more data out of the enterprise than any other workstation." By combining flow records, knowledge about the constituency, and NetFlow analysis tools, an experienced CND analyst can find a variety of potential intrusions without any other source of data. In fact, some mature SOCs focus as much or more analyst resources on flow analysis than with data coming out of their IDSes. NetFlow analysis is applicable to many stages of the cyber attack life cycle, whereas IDSes are traditionally oriented toward the reconnaissance and exploitation phases.

Flow records can be generated by a number of devices, including:

- Routers and switches; the official NetFlow record format actually originated from Cisco
- Some NIDS/NIPS, in addition to their normal stream of IDS alerts
- Some HIDS/HIPS, which may tie the flow to the OS process transmitting on the port in question, enriching the contextual quality of the data at the potential expense of extremely high volume, if widely deployed

- Software packages purpose-built for flow generation, collection, and analysis, such as SiLK [67], Argus [66], and S/GUIL [141].

SOCs often leverage purpose-built tools for their flow generation and collection needs because they can operate and control them directly, vice routers or switches. More important, however, the SOC can place flow collection devices where it needs them. This gives the SOC a tremendous advantage when analyzing how an advanced adversary moves laterally inside the network, something a border device would not see.

NetFlow tools are split into two parts, similar to a standard IDS deployment architecture:

1. One or more flow generation devices that monitor network traffic and output corresponding NetFlow records in real time
2. A central component that gathers flow records, stores them, and provides a set of tools for the analyst to interact with; analysis tools are typically either command-line based, leverage a Web interface, or both.

Some NetFlow tool suites can accept flow records generated by third-party systems such as routers or switches just like a native producer of flow records. Argus, for instance, can collect flows in Cisco's NetFlow version 5 and version 9 formats.

Some SIEM tools are also adequate at consuming and querying flow data—both in real time and retrospectively. With regard to NetFlow analysis, SIEM tools can process and alert on NetFlow records in real time, whereas, in traditional flow analysis, tools like Argus and SiLK are not usually used in this fashion.[1] That said, a healthy flow-based analytic framework will most likely leverage both real-time and retrospective analysis.

NetFlow is not without its limitations. Among these are:

- Just like any other network-based monitoring capability, a NetFlow sensor cannot generate records on traffic it does not see. Therefore, careful placement of NetFlow sensors across the constituency is key.
- Classic NetFlow records do not record anything about the content of network traffic above layer 4 of the TCP/IP network stack. Even if a flow is on port 80, that doesn't necessarily mean its contents are legitimate Web traffic.
- Under a heavy load, a NetFlow sensor is likely to resort to sampling a portion of the network traffic that passes by. If this occurs, records generated will contain incorrect packet and byte counts, thereby skewing derived statistics and potentially fouling downstream use cases.

---

1 There are some exceptions. The Network Situational Awareness (NetSA) Security Suite within System for Internet-Level Knowledge (SiLK) can do real-time processing of flow data [67].

- Like other network-monitoring capabilities, NetFlow analysis can be partially blinded by frequent use of NAT and proxy technologies throughout the network. For this reason, it is often prudent to collect flows from both sides of a proxy and/or the proxy logs themselves.
- NetFlow is sometimes used to perform analysis on encrypted connections because going deeper into the network stack is not useful. That said, the nature of the protocol must be carefully analyzed, because some combinations of tunneling and encryption can render NetFlow analysis marginally useful (e.g., the case with some uses of Virtual Routing and Forwarding/Generic Routing Encapsulation [VRF/GRE] and VPNs).
- Because they generate a record for each network traffic session, NetFlow records can dwarf most other data feeds collected by a SOC, especially if flows are collected from the end host.

When CND analysts look at an IDS alert, they see only something potentially bad about one packet in one network session. The flow records for the source and destination hosts involved in that IDS alert bring context to the analysts. What other hosts did the attacker interact with? Once the IDS alert fired, did the victim start making similar connections with yet other hosts, indicating a spreading infection? These questions can be answered with flow records and the tools necessary to query them. As we will discuss later, full-session capture can also support these use cases, but the beauty of NetFlow is that the amount of data needed to draw these conclusions is often vastly less, affording the analysts greater economy of data and speed.

For more information on NetFlow analysis, see [142] and Chapter 7 of [9].

8.2.4 Traffic Metadata

As we discuss in Section 8.4 and Section 9.2, there are a number of log sources that can be consumed by SIEM to spot potential intrusions. Three of the most popular are (1) firewall, (2) email proxy, and (3) Web proxy logs. These provide tremendously useful information such as email headers (sender, recipient, and subject) and HTTP headers (requested Uniform Resource Locator [URL], user agent, and referrer) for all traffic passing by. However, there are many situations where such logs are unavailable, hard to parse, or unreliable.

Fortunately, we have alternatives that take NetFlow one step deeper into the TCP/IP stack, providing analysts the same **traffic metadata** or "**superflows**" they would get from these other log sources. Metadata is roughly as voluminous as NetFlow, in terms of the number of records generated on a busy network link, but it can serve as an enhanced source of potential intrusion tip-offs. Consider a known set of websites that are hosting malware. This bad-URL list can be matched against incoming traffic metadata to look for potential infections. We can also collect metadata on DNS requests and replies, which can be used to look for malware beaconing and covert command and control of persistent malware. In fact, collecting metadata in the right places on the network allows us to be more selective in what is collected. Thereby, it presents less of a performance burden on both the SOC's collection systems and network services such as DNS servers or mail gateways.

Some tools such as Yet Another Flowmeter (YAF) and SiLK [67] and Bro IDS [65] provide robust metadata generation and analysis capabilities. Some NIDSes such as IBM Internet Security Systems (ISS) Proventia Intrusion Prevention [143] have series of signatures that also gather traffic metadata. These signatures are separate from attack signatures, and they may serve as an excellent stand-in for firewall or email gateway logs. Something to consider, however, is that many of these systems will log related pieces of metadata in separate events or log lines. In the case of email, sender address will be in one event, destination in another, subject in a third, and, perhaps, the file attachment name in a fourth. A correlation tool such as SIEM may be needed to piece together these disparate events.

8.2.5 Full Session Capture and Analysis

When we have a serious incident such as one that requires active response or legal action, we need concrete proof of what happened. While this can come from data pulled from the host, having a complete record of network traffic is often crucial. What is the content of this suspicious beaconing traffic? What was that user printing out at 2 a.m.? What was the full payload of this exploit, and what did this infected host download after it was infected? Full session capture can answer these questions, but we must be careful to scale our traffic collection and analysis platforms effectively.

Traffic capture is typically done on major perimeter connections in a sustained manner where an IDS (or other monitoring device) lives, and in an ad hoc manner near systems that are suspected of compromise, such as with adversarial engagements and other incidents. While we can filter out traffic we know we don't want recorded, due to volume, we still have a scalability challenge in all but the smallest deployments.

There are a number of tools that support full-session capture and analysis, the vast majority of which support input and output in binary libpcap format. PCAP is based on the original work that went into the libpcap libraries, on which capture tools such as tcpdump are based. There are many excellent sources [144], [68] for information on tcpdump and associated protocol analysis tools such as WireShark [69], so we won't cover their details here. At a high level, WireShark is a graphical user interface (GUI)-based utility that can record and display PCAP data in graphic format to the analyst. This allows the user to view how each OSI protocol layer is broken down for each packet. There is also a text-based version, tshark, that records and displays PCAP data, but without the fancy interface. Most SOCs leverage WireShark as a core-analyst tool because it is easy to use, decodes PCAP, sniffs low-bandwidth connections well, and is free.

The biggest challenge with full-session capture is obvious—volume. Consider an enterprise of tens of thousands of hosts that all go through one Internet connection—perhaps an OC-48 [145], which maxes out at more than two gigabits. At an average of 50 percent utilization, let's look at how much data we would collect in a 24-hour period:

2400 Mbit/s * 60 sec * 60 min * 24 hours * .5 utilization/8 bits per byte = 12.96 TB

At first glance, this does not seem like an insurmountable challenge. A single, moderately priced NAS can store far more than that, and 10 Gb network adapters are a commodity item. As we discuss in Section 9.2, Tier 2 analysts usually need at least 30 days of online PCAP for analysis and response purposes, preferably 60 days or more. That means we're looking at 30 * 12.96 TB = 388.8 TB. That's a very significant amount of data, even though we're looking at a moderately sized connection and modest retention.

While full-session collection and retention is not a trivial undertaking, there are several fairly straightforward ways to make it happen. Common open source utilities like tcpdump, careful OS and driver optimizations, and some careful scripting can be combined with commodity server hardware platforms to ingest PCAP at speeds well beyond one Gb per second. In most cases, it's easiest to connect a commodity server with large amounts of on onboard storage to a series of low-cost direct-attached storage devices (DASDs), each holding dozens of TBs in a few rack spaces. With careful attention paid to filtering out unnecessary traffic and compression of archived data, keeping a month or more of data online is not an unreasonable objective in most architectures. One of the biggest challenges for PCAP capture is packet loss, which is one of the top reasons for purchasing a capture platform specifically built to support packet capture at 10 Gb per second [146] and beyond. These platforms may be a better choice for SOCs that do not have the resources to construct and tune high-bandwidth data collection systems.

Specific traffic collection and analysis tools include NetWitness Decoder and Investigator [147], Solera Networks DeepSea [148], and AccessData SilentRunner [149]. These types of tools are almost always PCAP–interoperable but, in some cases, will actually record data in their own proprietary format. Whereas open source tools are free, many traffic-capture vendors license their products on a per-TB basis, making retention of large amounts of captured data quite expensive. The biggest advantage of these types of tools is that they both record data from high-bandwidth connections and present metadata about captured traffic to the analysts, allowing them to pivot and drill down very quickly from many days or weeks of traffic to what they need. Some SOCs will leverage FOSS tools to collect large amounts of PCAP cheaply and then run them through a commercial tool on an as-needed basis. This provides much the same usability but without the high price of a COTS tool.

There tend to be regular instances where the SOC suspects a particular constituent's system may be involved in an incident, requiring long-term collection of a narrow set of traffic. However, permanent emplacements of NetFlow or PCAP collection may provide little or no visibility into that end host. In such cases, it is helpful for the SOC to have a mobile platform for ad hoc PCAP capture, at a lower cost than the PCAP collection systems sitting off its main Internet gateway. Such a platform, perhaps a laptop system running Linux with a few TB of local storage, can be deployed on demand to an edge switch where the suspect host resides. These mobile platforms are instrumental in enabling a SOC to support long-term focused engagements assessing the adversary's TTPs.

8.2.6 The Case for Open Source and Virtualization

We have different options for which platform may serve as a basis for our network monitoring capabilities. Each option offers different performance, scalability, and economic advantages. As of the writing of this book, let's look at what you can get for $20,000 and one unit of rack space:

- Two CPUs with many hyperthreaded cores
- Several hundred GB of random access memory (RAM)
- Ten high-capacity spinning hard drives or, perhaps, ultra-fast solid state drives (SSD) that can be managed either through a hardware redundant array of independent disks (RAID) controller or a RAID-aware file system such as Z File System (ZFS) [150] or software RAID such as Linux mdadm [151]
- Four embedded Ethernet ports supporting one gigabit Ethernet (gigE) or potentially 10 gigE speeds
- Two or three Peripheral Component Interconnect express (PCIe) ports capable of supporting cards for:
  - A general-purpose graphics processing unit (GPGPU) such as Nvidia's Compute Unified Device Architecture (CUDA) [152] and OpenCL applications [153]
  - Multiple 10 gigE ports each
  - Specialized network capture cards from vendors such as Emulex [146] or Myricom [154], which can support 10gigE (and beyond) full PCAP with minimal packet loss
  - Host bus adapters for SAN or DASD connectivity.

That is a large amount of computing power in a small amount of space. Blade systems offer even higher density computing; however, storage and networking become more complicated, especially when considering networking monitoring applications. While the above specs obviously will become less impressive in the years following this book's writing, the point is that dedicating one server to one application is a thing of the past.

Consider enterprise-grade NIDS/NIPS products that can retail anywhere from $5,000 for a modest hardware appliance to over $100,000 for an appliance capable of multiple 10 gigE connections. Given these price differentials, let's look at our platform options as shown in Table 10.

Table 10. Commercial NIDS/NIPS Characteristics

Platform	Advantage	Disadvantage
Hardware appliance (specialized). Application-specific integrated circuits (ASICs) or field-programmable gate array (FPGA)-based hardware platforms optimized for NIDS/NIPS applications.	Special "secret sauce" hardware performs advanced protocol reassembly and analysis not possible with commodity hardware. Extremely high bandwidth; best-of-breed detection capabilities usually exceed any other point solution. Simplified device management.	Very high cost, limiting deployment to specialized use cases and customers with deep pockets. Some focus on niche markets; may not provide comprehensive coverage. Product is essentially a black box with limited insight by the end user. Specialized hardware design tends to take longer, meaning longer product update cycle.
Hardware appliance (commodity). Vendors package proprietary IDS software on commodity OS and hardware (typically Linux on x86). Most NIDS/NIPS products are sold in this form.	"Turnkey" deployment and simplified device management. Good performance out of the box, assuming normal tuning is performed. One high-end device can monitor multiple high-bandwidth network links.	Similar capabilities can be achieved through software IDS deployment on open market COTS hardware. Despite "appliance" name and premium price, little custom hardware is used, eroding value. Deployment across many sites is limited by device cost.
Software. Vendor ships product as package that must be implemented and tuned by customer.	Low(er) cost vice hardware appliances. Ability to deploy across inexpensive commodity hardware. Some limited user access to "under-the-hood" system components, at the risk of going outside vendor support. Potential for high-bandwidth applications, assuming ideal optimizations and high-end hardware.	Combination of IDS software and various hardware and OS platforms can complicate implementation and troubleshooting. Packet loss can become a problem if hardware choices and tuning are not carefully considered. Supporting fault-tolerant, robust in-line NIPS solutions can be a challenge. Few vendors still ship software NIDS.

Table 10. Commercial NIDS/NIPS Characteristics

Platform	Advantage	Disadvantage
Virtual appliance. Vendor ships propri-etary NIDS/NIPS in a virtual container with limited user access "under the hood."	Ease of deployment and manage-ment similar to hardware appliance platforms. Similar flexibility in choice of hardware and deployment as software-based NIDS/NIPS. Easily support monitoring within vir-tual environments.	Capability is comparatively new. Price premium may approach that of a hardware appliance. As with software NIPS, virtual NIPS deployments may still be challenging from a bandwidth and resiliency standpoint.

Network monitoring should enjoy the same benefits other areas of IT have gained from high-compute density. The commodity hardware platform can support at least six or eight monitoring points, with multiple detection technologies applied to each. In enterprises that have multiple points that require monitoring within close physical proximity, this is a com-pelling value. Some SOCs have engineering resources such that they can leverage world-class IDS technology for the bulk of their sensor fleet, without the premium price tag.

Consider what we discussed at the beginning of Section 8.2. We need collocated NIDS monitoring, NetFlow, and full PCAP collection. These tasks are easily accomplished by FOSS tools such as Snort, Argus (or a number of other tools such as SiLK), and tcpdump, with a bit of scripting to glue them together. In fact, disk input/output (I/O), rather than memory or CPU, may become a limiting factor for PCAP collection. Moreover, with the advent of network interface cards (NICs) purposely built for high-bandwidth IDS applica-tions, we may even be able to take on use cases previously left to high-priced Application Specific Integrated Circuit (ASIC) and FPGA-based platforms.

From a hardware perspective, integrating all of this equipment into one rack space is relatively straightforward. From a software perspective, integration is a bit more challeng-ing. There are two potential approaches.

The first approach is to run all monitoring software on top of the same OS, directly installed on the system as in "bare metal." In the case of multiple monitoring taps, each detection package will usually be running its own instance with a unique identifier, sepa-rate set of configuration files, and data output destination—be it physical disk, RAM disk, or syslog. The vast majority of SOCs that run open source software packages choose this approach. In the past, it was common to see only one physical hardware "pizza box" per monitoring tap point, keeping things very simple. Need another network monitored? Buy another server and put the same monitoring package on it. This approach is simple, cheap, and effective, but it depends on having sysadmins skilled in Linux or Berkeley Software Distribution (BSD), which is not the case for all SOCs.

Handling disparate detection packages can become problematic, especially when we want to bring a proprietary commercial IDS package into the mix, leading us to the second approach—virtualization. Also, the growing use of host virtualization technologies can make monitoring intra-virtual machine (VM) traffic more challenging; we may need a native, virtualized monitoring capability. For each monitoring point, we stand up a VM that contains one or more monitoring programs. If we wish to use multiple incompatible monitoring programs (e.g., both COTS and FOSS), they can be separated into different VMs. These can run on top of typical virtualization technology such as VMware ESXi/ESX Server [155] or Xen [156]. While adding some complexity, this approach adds a great deal of flexibility not found in a bare-metal OS install, as described in the following:

- Collapsing multiple, disparate sensors onto one system with fewer wasted resources
- Modular addition, removal, and transition from one monitoring technology to another
- Remote reconstitution, provisioning, and upgrades of guest OSes and their entire monitoring software suite
- Mix of multiple, incompatible monitoring technologies on the same interface, such as a FOSS solution running in one VM and a COTS virtual appliance running in another
- From a software and management perspective, each monitoring tool appears on a separate virtual host. When collecting events centrally, this makes sorting and correlating alerts for the analyst less ambiguous.

Figure 20 illustrates a high-level architecture of such a virtualized arrangement.

In this example, we have folded six disparate sensors onto one hardware platform—even greater consolidation is possible in practice. We have leveraged commodity hardware, free or cheap virtualization technologies, and FOSS tools to collapse our monitoring architecture and maximize the hardware resourced at our disposal. In order for SOCs to pursue this approach, they should pay careful attention to

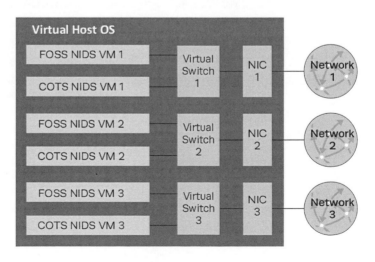

Figure 20. Virtualizing IDSes

optimizing their virtualization platform to their monitoring needs, being keenly attentive to any issues that may lead to packet loss. For example, using host virtualization of any sort may not work in very high bandwidth scenarios, due to its potential performance overhead.

Many constituencies are subject to specific threats and, thus, have a compelling need for custom signature support. From a cyber perspective, a SOC, better than anyone else, should know its own constituency and, therefore, be the best group to generate custom signatures. Snort is regarded as the *de facto* standard when it comes to writing custom IDS signatures, and many COTS IDS vendors offer support for Snort syntax. In fact, other FOSS IDS platforms such as Suricata [64], use signature syntax very similar to that of Snort.

When considering use of Snort-based custom signatures, the SOC has a few things to consider:

- What proportion of the sensor fleet will need to fly these custom signatures?
- How many custom signatures are needed—a few dozen or several hundred? Some IDSes pay a higher performance penalty for custom signatures than native ones.
- How does the COTS NIDS support Snort signature implementation? Is it actually running Snort, and, if so, how old is it? If it's emulated, how good is the emulation and are all Snort preprocessors supported?
- If the SOC already has a COTS NIDS with custom signature support, does that custom syntax support its needs? Is it willing to spend time translating Snort signatures provided by other SOCs into this custom signature syntax?

Some SOCs, which favor widespread use of "bleeding edge" Snort signatures, may choose to use native versions of Snort. Other SOCs, which don't have strong UNIX/Linux expertise, may choose to go down the commercial path—either with the COTS version of Snort, Sourcefire, or another vendor that implements Snort signature support. SOCs can avoid running an extra Snort sensor fleet if they feel their COTS NIDS has sufficient Snort signature support. Having more than one IDS engine can give a SOC extra options when facing an elusive or targeted threat.

Virtualization aside, let's summarize the some typical differences between best-of-breed FOSS and COTS NIDS/NIPS platforms. (See Table 11.)

Table 11. COTS Versus FOSS for Network Monitoring

Characteristic	COTS	FOSS
Code base	Closed	Open
Signatures	Closed. This can be very frustrating when an analyst wants to understand why an event fired; the number and robustness of custom signatures are sometimes very limited.	Open. Being able to understand and write signatures with the same fidelity for what ships with the product is an absolutely key feature; typically thousands of custom signatures can be flown if necessary.
Protocol detection engine robustness	Usually very good. A major value-added as it is a vendor's "secret sauce."	Also can be very good, but is up to the community to keep pace with evolving threats and protocols.
Predisposition for false positives	Varies widely depending on the individual signature and detection engine.	Varies widely depending on the individual signature and detection engine.
Availability of signatures for critical new vulnerabilities	Within days	Often within hours
Bandwidth	Capable of handling multiple gigE taps; more expensive products advertise 10 gigE.	Most solutions can handle gigE monitoring with little issue on modest hardware; scaling to 10 gigE and beyond typically requires attention to hardware and software optimization [157].
Management complexity	Almost always point-and-click, which makes training new staff straightforward, but some systems can be deceivingly hard to manage due to several layers of complexity; some COTS solutions become difficult to manage with fleets of hundreds of sensors.	For the novice, this can be a daunting task; experienced Linux/UNIX sysadmins can usually automate management of large fleets of sensors with very little sustained labor by leveraging tools native to the UNIX environment.
Overall suitability for in-line prevention (NIPS) use	Depends on the product implementation; usually very good to excellent.	Manual configuration of commodity hardware and OSes can make this problematic if the system is not built from the ground up to be fault tolerant.

Table 11. COTS Versus FOSS for Network Monitoring

Characteristic	COTS	FOSS
Combined cost of software and hardware to implement full solution	Moderate to very expensive, depending on bandwidth requirements; expect yearly maintenance costs to be around 20–25 percent of initial acquisition.	Assuming the availability of a few talented Linux/UNIX administrators, this can be very cheap, especially with deployments of 50+ sensors; the only outyear cost is hardware maintenance.

Both FOSS and COTS IDS platforms offer compelling features and drawbacks. Mature SOCs leverage a measured combination of both in their monitoring architectures.

8.2.7 Malware Detonation and Analysis

If there was a "killer app" for NIDS/NIPS, it would probably be detecting and blocking direct, network-based attacks such as buffer overflows. These attacks faded in the late 2000s. Vendors improved their protection against these attacks against code that runs primarily on servers and applications directly exposed to the Internet. As a result, adversaries shifted their tactics to exploiting client weaknesses through phishing and pharming attacks. In these newer client-side attacks, users are tricked into downloading malicious content from websites or emails, respectively. This strategy works for two reasons: (1) there were still a large number of vulnerabilities in programs running on end hosts, and (2) attackers could easily target those vulnerabilities by placing malicious content on websites or send them to potential victims through email.

While NIDS/NIPS can provide some limited help here, this set of attacks is usually beyond IDS's capabilities. As a Portable Document Format (PDF) or Word document passes by on the wire, the IDS sees a stream of encoded binary information. The NIDS doesn't usually have time to assemble the file and, therefore, doesn't understand what it does when it runs.

Enter a new breed of malware "detonation" or "sandboxing" products such as FireEye AX [81], Norman Shark [158], and ThreatTrack TreatAnalyzer [80] that delve deeper into files pulled from network traffic or from end systems. These products are sometimes also known as "next generation AV" and blur the line with other product offerings [159]. Their main purpose is to accept potentially malicious files, "detonate" them in an artificial environment, and observe the behavior of the files at execution time.

Malicious files will usually behave in suspicious ways like making privileged system calls, beaconing out for command and control, or downloading additional malicious packages. These malware detonation systems will look for this sort of activity, but without signatures that define a specific attack or vulnerability. As a result, they are uniquely tuned to ongoing detection of zero-days and specially crafted malware. While the attack vectors may change, the outcome doesn't, and that's the system's focus. Even if malware is obfuscated or packed inside a binary file, the detonation chamber should still be able to notice that it is acting in an anomalous or malicious way—something an IDS is not capable of.

Content detonation systems come principally in two "flavors." The first type accepts file uploads by users in an "offline" manner. This is particularly useful for Tier 2 incident analysts and malware analysts who need an on-demand capability that can provide quick details on whether a file is likely to be malicious or not. This capability can automate several hours of manually intensive malware reverse engineering. Best-of-breed products will provide specifics on system calls, network connections made, and files dropped. Some malware will actually try and detect whether it is running inside a virtual analysis environment and then change its behavior—a good content detonation system should also detect this.

The second type is a device that can scan network traffic (usually Web or email) in real time, pull out malicious files, and detonate them fast enough to actually block the malicious content from reaching the victim user or system. This kind of device is targeted toward use cases where a SOC actually wants to detect or block client-side attacks in real time, thereby catching a lot of the "ankle-biter" malware traversing the constituency perimeter. Some systems can also take the "bad" files they've seen and automatically produce IDS signatures that can be leveraged to see whether the same file popped up elsewhere on the network.

At the time of this book's writing, these devices are in vogue. They are being rapidly deployed in large enterprises and the marketplace is expanding. That said, there are some limitations and cautions regarding content detonation devices that should be recognized:

- Many malware authors recognize that malware reverse engineers and content detonation devices will attempt to run their malware in a VM. As a result, the malware is built to be "VM-aware" and simply will not execute when in a VM, resulting in a false negative for VM-based detonation chambers.
- Some malware requires user interaction before it will execute. The detonation chamber presumably has no user to click buttons in a dialogue box that will then trigger the malware to run, thereby resulting in a false negative. Building and configuring a content detonation system to cope with this can be a challenge.
- Some exploits such as heap sprays require very specific arrangements of an OS's components in memory. Because malware detonation systems try to fit as many VMs as possible in a modest platform, there may not be sufficient memory to fully simulate

what would run on a real end host. A VM may only have one GB of memory allocated, whereas a heap spray may require at least four GB to execute. Under such conditions, a false negative would result.

- A lot of malware is less than perfect and may not successfully exploit a host every time it is executed. A content detonation device most likely only has one opportunity to detonate a file that is potentially malicious. As a result, the malware may not be caught in the single time it is run, resulting in a false negative.

- Some malware may wait a certain amount of time before executing. The content detonation device will time out after a certain number of CPU cycles or seconds. Some malware authors will specifically write their malware to delay execution in order to avoid detection by AV engines or content detonation devices.

- The content detonation device will open files within some sort of VM or sandbox that is meant to match common corporate desktop configurations. However, these configurations may diverge significantly from what is actually being used in the constituency. The operators of the content detonation device should make sure their VMs/sandboxes match the OS, browser, browser plug-in, and application revision, service pack, and patch level as used on their corporate desktop baseline. Failure to do so may result in either false positives or false negatives.

- If used in in-line blocking mode, the content detonation device may serve as an additional point of failure in email or Web content delivery. If the constituency uses redundant mail or Web gateway devices, the SOC may consider also dedicating a content detonation device to each gateway device, thereby preserving the same level of redundancy but making deployment more expensive. At a minimum, the content detonation device should fail "open," allowing traffic to pass if it fails.

- A given content detonation device can execute only a certain number of files in a given period of time. Use on very high-speed links can be problematic because some files may never get executed, or the devices may run into bandwidth limitations. SOCs should work with vendors to ensure products are properly sized for their constituency size and that load-balancing techniques are used where necessary.

- Content detonation devices will likely open a SOC's eyes to the malware that the NIDS, HIDS, AV, and content-filtering devices never picked up, potentially alarming some staff and seniors. That said, the malware reports generated can only be correctly interpreted by someone with experience in malware reverse engineering.

- SA provided by a content detonation device will make it very clear how many malware hits a constituency had in a given day or week. Without additional context of whether these were mitigated or blocked, great alarm could result, even though concern is not warranted.

Malware detonation and analysis systems are powerful in large part due to their elegant concept and design. Despite their limitations, their use should be carefully considered by almost any SOC.

8.2.8 Honeypots

Whereas content detonation and analysis systems simulate an environment allowing the rapid discovery of malware, some SOCs may wish to mock up a full-fledged host or network environment in order to find and study adversaries such as would-be spammers. **Honeypots** are a set of computers set up by network defenders with the primary intent of luring in attacks and trapping them in a highly instrumented environment [160]. A comprehensive discussion of honeypots is beyond the scope of this book, but more information on them can be found in [7] and [161]. Generally, only the most well-resourced SOCs will deploy honeypots. We mention them here because they are a well-known technique leveraging many detection tools discussed in this section. They can result in increased intelligence of attacker behavior, allowing a SOC to better instrument its defensive capabilities.

8.2.9 The Fate of the Network Intrusion Detection System

When many large SOCs were first stood up in the late 1990s, NIDSes usually dominated any discussion of what it took to perform intrusion monitoring. Life was comparatively simple—attacks propagated across the network and a NIDS was the way the SOC detected them.

In 2003, Gartner, Inc., declared NIDSes dead [162]. They made the point that NIDSes' prodigious quantity of false positives renders them not worth the trouble—a conclusion that some SOC analysts have also arrived at. After all, a NIDS doesn't *do* anything other than produce lots of data that some analysts find to be of questionable value. It follows that we should, instead, focus on our attention on NIPSes, which actually block attacks. At the end of the day, though, NIPSs are built on the same concepts as NIDSes, while their producers are more careful about keeping false positives under control with, presumably, more robust detection engines and better written signatures.

Today, NIDS technologies are *one* among several technologies that a SOC will leverage to find potential malicious activity within the constituency. When leveraging LM or SIEM, a good feed of security-relevant logs may be just as valid a source of intrusion tip-offs as a purpose-built NIDS. More important, exploits executed across the network (most notably remotely exploitable buffer overflows) no longer constitute the overwhelming majority of initial attack vectors. Client-side attacks through phishing and pharming have become far more prevalent, giving way to the content detonation and analysis devices we talked about in the previous section. This, in addition to application logic attacks such as SQL Injection, requires reconstruction of protocols and behavior at layer 7 and above. As a result, signature-based methods *by themselves* (e.g., AV and most NIDSes) are no longer considered sufficient for finding attacks and defending a network.

Does this mean the NIDS is truly dead? Not necessarily. It is a near-certainty that there will always be a network-based attack detection and response mechanism *of some sort* used by the SOC. Network-based monitoring technologies are usually the most cost-efficient and simplest means by which SOCs can gain visibility and attack detection coverage for a given enclave or network, especially in cases where they have no other visibility. Whether it's a traditional, signature-based NIDS or something else is a separate issue. We clearly recognize a NIDS as just one tool in a larger suite that is, unfortunately, growing in complexity and cost. This is the reality of defending the modern enterprise.

For more information on the history of intrusion detection, see Appendix B of [9].

8.3 Host Monitoring and Defense

Network sensors have many virtues—one sensor can give us SA and tip-offs for potential incidents across thousands of systems. But their insight is only as deep as what can be seen in network traffic. To complement our visibility, it is also useful to instrument the end hosts with a variety of detection and blocking techniques.

Details suggesting, confirming, and elaborating on the presence and penetration of the adversary can best be found through monitoring and analysis of the end hosts' content. Moreover, incident response often involves touch labor on end hosts, a process that can take hours, days, weeks, or even longer. Consider an enterprise with well-instrumented desktops and servers that not only provide tip-offs and comprehensive context about an intrusion but also enable automated response actions. Many mature SOCs are compelled to pay as much attention to instrumenting end hosts as they are to the network.

This section encompasses the scope of host sensor instrumentation used by the SOC to detect, analyze, understand, monitor, and prevent security incidents *on the end host*. These tools take the form of a software agent installed on the host that observes local host activity. Similar to a NIDS, host tools are controlled and monitored by a central management system. Whereas NIDS/NIPS deployments can comprise dozens or hundreds of sensors, a SOC may have a host IDS/IPS deployed on every end host. In the case of a large enterprise, this could comprise hundreds of thousands of sensors. With such a wealth of data and the heterogeneous nature of the IT enterprise, scalability is a challenge.

There are a number of capabilities we wish to bring to bear at the host level. While there are some niche products that only focus on doing one thing really well, more typically we see one product combine multiple capabilities:

- AV/antispyware
- Intrusion detection and prevention

- Application blacklisting and whitelisting
- Configuration tracking
- Network access control
- Host-based firewall
- IP loss prevention
- User activity monitoring
- Remote incident and forensics support.

Products that try to cover many or most of these features can be regarded as a Swiss Army knife—useful for many jobs when it's all you have but sometimes inferior at any one task. Despite this, many SOCs feel compelled to go for the Swiss Army knife approach, due to budgetary and resource constraints. Also, integration into a diverse IT environment is a challenge, and each tool must be regularly tuned. Tools from different vendors have been known to recognize each other as malware, making coexistence of specialized tools a challenge.

Almost every tool will leverage a set of observables contained in system memory, CPU, disk, peripheral contents, or a combination of these. The differences among tools lie in their targeted features and the techniques (e.g., "secret sauce") that leverage various observables to support said features. As a result, we will begin our discussion by considering these "observables" in some detail. Following that, we will discuss how they are leveraged for each major capability implemented in best-of-breed host-based monitoring. Finally, we will finish with key considerations for use of these tools in practice.

8.3.1 Observables and Perspective

When we have a presence on the end host, there are a multitude of possibilities for what we can learn about what the actors are doing. These observables can be gathered either on an ongoing basis or on demand and synthesized in many ways. Let's start with the building blocks:

From mounted file systems and any other storage:

- OS version, installed service pack(s), and patch level
- Installed applications
- Resident files, their modification times, ownership, security permissions, contents, and summary data such as size and cryptographic hash value [163]
- File system "slack space" containing deleted files and recycle bin/trash can contents
- Contents of the entire physical disk such as a bit-by-bit image
- OS and application logs
- OS and application configuration data such as the contents of the Windows registry hive

- Browser history, cache, cookies, and settings.

From system memory and processor(s):

- Application process identification number (PID), creation time, executable path, execution syntax with arguments, name, user whose privileges it is running under (user context), CPU time used, and priority
- Actions and behavior taken by running processes and threads, such as execution behavior and system calls
- RAM contents and memory map
- Clipboard contents
- Contents and disposition of CPU registers and cache
- Logged-in users or applications acting with privileges of a remote user such as with a database or custom application.

From attached devices and system I/O:

- Network flow (sometimes known as "host flow") data, possibly including enrichments that tie process name to the ports and connections it has open
- Content of network data traffic
- User keystrokes
- Actions from other input devices such as mice, touch pads, or touch screens
- Screenshots
- Connected devices, potentially including details such as device type, driver info, serial number/ID, system resources, addressable storage or memory (if applicable), and insertion/remove events.

In order to paint a complete picture of what is happening on the host, it is usually necessary to examine all three of these elements (on disk, in memory, and attached device I/O). For instance, focusing on just the local file system will blind an analyst to malware operating exclusively in memory. The host monitoring package must also implement its data collection in a manner that doesn't obligate a large portion of system resources, especially CPU and RAM.

Depending on where in the system we sit, the data we're trying to gather can vary widely. We have the options described in the following paragraphs.

Most host monitoring tools reside on disk and are run in memory like any other program. In this scenario, the tool must verify that when it starts its code has not been compromised. By leaving a permanent presence on disk, it can run automatically each time on startup, but this makes it easier for malware to recognize its presence and circumvent detection.

Some host monitoring tools can run entirely in memory, with little or no on-disk footprint. This technique provides some sort of installation or injection of code into the OS at boot time or shortly thereafter. This closes off some opportunities malware has to undermine the host monitoring tool but introduces added complexity from a management and distribution perspective.

In virtualized environments, it is possible for the host monitoring tool to live at the virtualization layer, using introspection techniques to "see into" virtualized guests but without the guest being aware of the monitoring tool's presence. These techniques, while present in academia for some time, are still nascent in the commercial marketplace as of this book's writing. Some products take a hybrid approach, with the main monitoring framework sitting at the parent host layer and tools reporting observed behavior from within the VM.

When the monitoring package must reside on the host being monitored, it usually resides in "ring-0" [164] where the OS runs with system-level privileges. This, unfortunately, places it on equal footing with malware that obtains the same system-level privileges. This results in something of an arms race between malware authors and security vendors to defeat or circumvent each other on the host. Furthermore, malware that resides at the firmware, BIOS, or hardware has the advantage at defeating these monitoring packages, at least in part, as famously claimed in [165].

Some more esoteric host monitoring approaches leverage permanent storage or a root of trust at the hardware level. Rootkit detection, for instance, could be driven by specialized monitoring tools implemented as a peripheral component interconnect card in a system [166] but comes at prohibitive cost. There are also ways of using the trusted platform module (TPM) [167] in modern Intel architecture systems as a root of trust, ensuring both the OS and other components have not been compromised. This work, however, is still mostly in the research phase and small pilots as of this book's writing.

When considering any sort of host-based monitoring package, the CND architect is well advised to consider tools that are able, first and foremost, to defend themselves against the attacks they attempt to detect or prevent. For instance, a tool that detects rootkits is useless if it runs with user-level privileges and therefore is easily subverted by rootkits running with system-level privileges [168]. In this case, the monitoring package would likely be fooled into seeing the system's content manipulated by what the rootkit wants it to see.

8.3.2 Antivirus and Antispyware

The topic of AV tools should be familiar to the readers. In short, we have a program that inspects file system and memory contents, leveraging a large signature pool and some heuristics to find known malware or known malware techniques. Just as with signature-based IDSes, they may be circumvented [169]. They are not a *complete* defensive tool but are, arguably, still *relevant* in the context of Windows systems that are impacted by the vast majority of viruses in existence. Considering their limitations, a common criticism of AV tools is that their system resource utilization, RAM footprint, and regular disk scans outweigh their diminishing benefits.

Antispyware capabilities are often included in most AV suites. They add to their malware detection capabilities by also looking at Web browser specifics such as stored cookies, embedded Web page content, browser extensions, and stored cache. With these features, AV packages add some more modern value, especially for users wishing to rid their systems of some malware that comes from regular Web surfing.

There has been significant attention paid to the fact that AV tools only detect a small percentage of all malware, and, of the malware they do detect, it is almost entirely malware running on Windows/Intel-based platforms. Mandiant, a recognized vendor of host-based incident response tools, puts the detection rate of AV tools at 24 percent, when considering "APT" malware [170]. Other sources have quoted percentages of anywhere from 15 to 40 percent [171], [172]. In AV's defense, 15 to 40 percent is better than zero percent, and, at the very least, AV can provide indicators that a host is infected. One thing to keep in mind, though, is that an AV tool will often report on a virus and report the system as "cleaned" when, in fact, it has only picked up on a portion of infection, leaving the adversary's other tools and persistence on the system to continue unabated.

Mandated use of AV in the corporate and government environment is still overwhelmingly common. Operationally speaking, use of AV on the Windows desktop is generally judged as worthwhile, albeit marginally [173]. AV on non-Windows platforms such as Apple, Linux, and UNIX is regarded as unnecessary by some network defenders, despite the fact that many organizations issue blanket "must deploy" AV policies for every desktop and server in the enterprise. Nearly every large enterprise has a mandated host-based monitoring and prevention tool of some sort, and, in this regard, AV is considered the lowest common denominator. As we mentioned above, if a SOC deploys any tools other than AV, those tools may flag each other as malware or, in the most extreme cases, crash their host due to conflicts. SOCs must pay careful attention to integration prior to deployment.

For more information on the detection rates and other comparisons between popular AV products, see [174].

8.3.3 Host Intrusion Detection System/Host Intrusion Prevention System

Given our discussion of IDS from Section 2.7 and Section 8.2, one can easily recognize the possibilities for detection and prevention of malicious activity at the host. Indeed, HIDS/

HIPS tools generally leverage both signature-based and behavior-based detection techniques. Whereas the cornerstone of AV is a library of signatures for millions of malware strains, the cornerstone of HIDS/HIPS is catching deviations from what is considered normal OS and application behavior.

HIDS/HIPS tools typically rely on a set of behavior profiles for everything running on a host. These behavior profiles can be built through periodic or continual "learning" or "training" of how hosts operate under what is presumed to be "normal" behavior. When hosts stray from this behavior baseline, an alert will fire, and, if set to prevent, a HIPS will actually block the activity. Imagine, for instance, if Microsoft Word suddenly started writing various Windows registry keys or opened communication to a remote server over a nonstandard networking protocol. A HIPS suite should notice and block this.

Drawing another parallel to network-based IDS, a HIDS can be circumvented by advanced strains of malware. One classic technique on the Windows platform attacked vendors' common reliance on how they examined key system components. For instance, a HIDS would pay close attention to any file modifications to c:\windows\system by monitoring the Windows file system handlers. Malware authors circumvented this by remapping this directory with an arbitrary name or by issuing direct I/O to the files contained within its directories, thereby avoiding detection. While this specific attack was recognized and resolved years ago by best-of-breed vendors, it illustrates the cat-and-mouse game that hinders most HIDS and HIPS tools to this day.

Out of all the host monitoring suite components, HIDS/HIPS are often regarded as the most problematic for two reasons. First, they are deeply integrated with the host OS. In large enterprises with different system baselines, HIDS/HIPS packages can frequently cause nontrivial conflicts with other running components, so a certain level of maintenance and debugging before full deployment is always necessary. Second, it is always appealing to deploy a HIPS to a large portion of hosts, possibly all servers and in some cases all desktops. This can be a challenge because any missteps with HIDS/HIPS signature or profile tuning can easily lead to widespread service interruptions and an influx of help desk calls. While these issues can crop up with any host-based monitoring tool, they seem to be most prevalent with HIPS.

Popular HIDS/HIPS suites include IBM Security Server Protection [175], McAfee Host Intrusion Prevention [176], Symantec Endpoint Protection [177], and Sophos HIPS [178].

8.3.4 Application Blacklisting and Whitelisting

One offshoot of HIDS/HIPS and AV is the ability to perform more proactive control over what can and cannot run on the end host. Application blacklisting is a technique whereby an OS module or protection agent blocks unwanted processes running on the end host [179].

It does this by monitoring for processes that match a certain set of criteria such as process names, a code's MD5 hash [163], or whether the code has been signed by a trusted root certificate authority. Blacklisting is akin to a "default allow" policy where only certain applications are prevented from running.

Application whitelisting is similar to application blacklisting, but instead of having a "default allow" policy, it uses a default deny approach. Sysadmins must define which programs are authorized to run on which systems; all others are blocked from running either by the OS or by the blacklisting/whitelisting client.

Application whitelisting and blacklisting may be built into some OSes (e.g., AppLocker in Windows 7 and Windows Server 2008 [180]). It is also often part of a HIDS/HIPS suite (e.g., Bit9's Application Whitelisting [181] and McAfee's Application Control [182]). As is the case with both AV and HIDS/HIPS suites, some enterprise-grade implementations of application blacklisting and whitelisting are known to be circumvented by a number of techniques [183], but are nonetheless recognized as a means to raise the cost of successful exploitation [184]. Vendors have made vigorous efforts to close the holes in their protection schemes, but, as with any other signature-based protection scheme, there is a certain "arms race" between white hats and black hats.

SOCs wishing to pursue application whitelisting or blacklisting technologies should consider the additional management overhead involved in tracking allowed or denied applications on the enterprise baseline. In order to implement whitelisting, all monitored hosts must adhere to a known OS and application baseline, and the SOC must continually maintain absolute consistency with that baseline (lest a whitelisting tool stop a legitimate application or service from running). This can be especially problematic with a complex enterprise baseline or decentralized IT administration. As a result, many SOCs do not leverage whitelisting, due to the large time investment it entails.

8.3.5 Configuration Tracking, Attestation, and Enforcement

Strong configuration management is universally regarded as a key enabler to a strong defensive posture for the enterprise. A nexus of this can be found with configuration monitoring at the desktop and server.

Tools exist that passively track and attest to system configuration, such as the classic open source version of Tripwire [185]. Traditionally, Tripwire is used on UNIX hosts to detect changes to key configuration files. While this can support proactive CM and change tracking, it can also be used to alert on changes that may be an indicator of malware or a malicious user. Tools that report on system configuration settings (and changes to them) can also be used to propagate configuration changes to systems. The commercial version of Tripwire [186], which is available for a variety of UNIX and non-UNIX platforms, is one such example. Changes that are detected in monitored files and settings can be reversed by the administrator. Aside from system settings, a focus area for these tools is checking the status of system patches. Whereas an enterprise may leverage one tool to push patches to the desktop and server, one way to cross-check compliance is to use a different tool that verifies the patches were successfully applied.

Taken to their logical conclusion, configuration tracking tools can be combined with automation at the network infrastructure layer to shape services provided to systems (depending on their configuration compliance status). This is a key feature of **network access control** (NAC) systems sold by several companies, including PacketFence [187], McAfee [188], Juniper [189], and Microsoft [190].

Imagine a constituency with a large number of VPN users who may not connect to the corporate network for weeks or months at a time. When these users connect, their systems are likely to be significantly out-of-date, presenting a risk to other constituency systems. Once logged in via VPN, NAC can be used to limit those systems' access to network resources such as patch servers until their systems are brought up to date. The NAC client installed on the end system will examine specified system attributes such as patch level and report those to the NAC server, which grants network access to the end hosts.

There are a number of operational considerations to deploying NAC. For instance, are IT administrators willing to keep key executives from their email because their patches are a few weeks out of date? Can the help desk support increased incident load from users who experience issues with their NAC client unsuccessfully recognizing their patches are up-to-date? Have network administrators tried enforcing network switch port security [191], and, if so, did they have the resources to keep up with constant changes to those IT assets that were plugged into the network? Finally, while these tools can be used to push configuration changes, they are generally not regarded as a comprehensive end-system management suite or patch distribution system.

Regardless of the tool or technique used, skilled adversaries will be keenly interested in ensuring that CND analysts are not alerted to any changes resulting from their presence on constituency hosts. To that end, they will go to great lengths leveraging tools such as rootkits to shield changes to key system files that would trigger a Tripwire or NAC alert. In some cases, these tools could be attacked directly and made to provide false information to their upstream management servers. This is the case for any host-based monitoring tool but is most acute with configuration tracking and HIPS suites.

While all but the simplest host monitoring tools will leverage a variety of internal checks to guard against compromise, none are foolproof when malware stands on an equal footing (ring 0) with monitoring packages. To this end, the TPM [167] may be used [192] as a root of trust for host configuration attestation.

8.3.6 Firewall

Although firewalls are most widely recognized as appliances that filter traffic crossing between two or more networks, host-based network traffic filtering capabilities can be found in virtually all popular varieties of UNIX and Linux and will usually be included in most HIDS/HIPS suites. In UNIX and Linux environments, host-based firewalls are primarily used to limit external systems' (and users') access to sensitive services such as remote management tools (e.g., SSH). There are many host-based firewall packages—varying by particular flavors of UNIX—including IP tables [193] and Packet Filter (PF) [194].

Desktop firewalls, as they are commonly known, are generally used to augment rules already in place on network firewalls and gateways. These can be used to further hamper the spread of network-borne malware, especially in the presence of a pressing or elevated threat. It should also be noted that many intruder tools use legitimate ports and protocols, rendering less sophisticated desktop firewalls of limited use in countering the same malware. Firewalls on the desktop are certainly not a replacement for enterprise-grade counterparts that sit at network gateways and are almost always used as a supplement to them and not a replacement. Symantec [195], CheckPoint [196], and McAfee [197] all provide host-based firewalls with their HIPS suites.

8.3.7 Intellectual Property Loss Prevention

With the increasing prevalence of encrypted network protocols and high-capacity removable media such as universal serial bus (USB) "thumb" drives, there is significant concern about the exfiltration of sensitive data from the enterprise. This can include anything from sending sensitive documents over personal email to downloading HR data to a thumb drive. The host is often the only place where we can expect to clearly see this activity (e.g., through network traffic and system call observables). As a result, many of the enterprise host monitoring packages listed in this section include functionality that will scan and report on data transferred to local removable media as well as website and email postings (known as intellectual property or data loss prevention [DLP]). Some intellectual property loss prevention packages can also be used to block or limit user access to removable media, enhancing functionality already present in Windows domain GPOs [198].

8.3.8 User Activity Monitoring

In some enterprises, there is a significant concern over the actions of a large portion of the user populace. Many organizations must follow a policy of "trust but verify," whereby users are given latitude to perform their job functions, but their actions are heavily monitored. These may include any organizations that handle large amounts of sensitive or high-value data, such as defense, intelligence, finance, and some areas of industry. In such cases, a security, counterintelligence, or intellectual property loss prevention shop may require full-scope user activity monitoring. Typically, these capabilities involve comprehensive capture of user activity on desktops, to the point where users' actions can be monitored in real time or replayed with screenshots and keystrokes. The efficacy and legal issues surrounding use of such software is outside the scope of this book.

However, as we can see from the observables mentioned above, the host is obviously the right place to perform such monitoring. Such packages will usually encompass functionality also found in the intellectual property loss prevention tools mentioned above. Commercial examples of insider threat monitoring packages include Raytheon SureView [199] and ObserveIT [200].

8.3.9 Incident and Forensics Support

So far, the vast majority of our discussion with host monitoring has been focused on the left-hand portion of the cyber attack life cycle, whereas we are interested in the entire attack life cycle. When one box gets hacked or "popped," the SOC must answer questions including: have any others been compromised? How do we know for sure if we never received an IDS or AV alert from any other hosts? What if a compromise was discovered through NetFlow analysis or a user calling in? Moreover, what do we do considering our SOC is in Arizona and the compromised system is in Morocco?

In the latter half of the 2000s, several vendors have brought to market a series of products meant to aid SOCs in rapidly evaluating the impact and spread of compromises in the constituency. These products include Mandiant for Intelligent Response [201], AccessData CIRT [202], HBGary Responder Pro [203], and Guidance EnCase Enterprise [204]. These tools are primarily designed to leverage the full scope of observables described at the beginning of Section 8.3, sometimes known in this context as "host telemetry," in support of ad hoc intrusion analysis. In addition, many of these tools can pull entire images of RAM or disk for remote analysis.

For instance, imagine that a SOC analyst has a compromised system to deal with. Quick memory and disk analysis has revealed a set of programs that appears to be suspect. The aforementioned tools will allow the analyst to scan other systems in the constituency for evidence of the same files, perhaps through memory map analysis or file hash scanning on disk. Some tools, particularly EnCase Enterprise, will actually remotely pull a partial or full image of system contents for analysis.

Or, let's assume the enterprise has a fairly standard desktop and server system build. This suggests that the processes running on each host should be relatively consistent, barring specialized applications or one-off system builds. Using one of these tools, the analyst can query host telemetry for programs or behavior occurring on only a small number of hosts—an indication that they may be compromised or at least subject to nonstandard configuration practices.

Although there are certainly scalability, performance, and bandwidth implications in using these tools, they can enable a variety of incident response options not otherwise available. Timelines for incident analysis can be shortened from weeks and months to minutes and hours. SOCs that deal with multiple system compromises every week, especially at remote locations, are compelled to leverage remote incident response tools as part of the monitoring and response architecture. These tools, perhaps more than any other, allow a SOC to stay within the decision cycle of the adversary, even with a large, geographically dispersed enterprise.

8.3.10 In Practice

Having a monitoring and response capability on constituency hosts can be an indispensable tool when used effectively. However, there are a number of cautions the SOC should keep in mind when considering deployment and use of various host-based tools discussed in Section 8.3. As a result, we offer the following tips for success:

- **Host-based monitoring and protection is not a panacea.**

  Some SOCs are put into a position of overselling the capability to upper management in order to get the capability funded and/or mandated. Managing expectations of IT seniors when it comes to any defensive tool is a challenge; with HIPS, this problem seems to be especially acute. Many seniors get the impression that with the deployment of this widget called "HIPS" our hosts are "protected" and "we're good." Of course, this is not the case. As usual, the SOC is advised to pursue a careful strategy for approaching upper management when deploying any pervasive tool such as HIDS/HIPS.

- **Commercial tools are primarily aimed at the Windows desktop, and secondarily the Windows server environment.**

  In many cases, they are only marginally relevant in a UNIX, Linux, MacOS, mainframe, or embedded appliance context. For instance, on a Linux server, a combination of commonsense system lockdown, IPchains, log aggregation, and Tripwire may be more than sufficient.

  SOCs are encouraged to work with constituency executives and sysadmins to pursue a commonsense approach to ensuring comprehensive host monitoring coverage, while ensuring platform and threat relevancy. Most SOCs have success with making their host monitoring packages part of the constituency server and desktop baseline package, engineered and deployed by the respective groups in IT engineering and operations.

Experience suggests, however, that systems in greatest need of robust monitoring and prevention tools are often the most fragile or most antiquated (such as mainframe platforms) and, therefore, poor candidates for most COTS host monitoring packages.

- **Aggressive host monitoring expands the responsibilities of the SOC.**
Setting the signature and protection policies of these is the SOC's job because it is a defensive capability. System management, on the other hand, is the job of IT operations, of which SOC may or may not be a member. If the SOC creates and pushes a sensor policy that interrupts or degrades constituency services, it will likely be hit very hard politically; repeated mistakes will often burn a lot of the SOC's political capital and/or drive IT executives to reconsider continued use of said capabilities.

When deploying active prevention capabilities on the host, SOCs are advised to pursue a formal CONOPS cosigned by IT operations seniors that gives the SOC timely control over signature policies and other monitoring changes, but keeps IT operations and the help desk informed and involved. The SOC should carefully manage the resources it devotes to smooth operation of host monitoring suites and integration with constituency desktop and server baseline(s). Unfortunately, this may compel the SOC to not pursue HIDS/HIPS capabilities for non-mission–critical systems.

- **Consistent desktop and server baselines and centralized management are usually prerequisites to use of a host-based monitoring tool.**
Deploying and upgrading these tools requires administrative access on monitored systems. In the case of a Windows domain, this is relatively straightforward. In cases where this is not true (such as a very large or fragmented network), deployment and management will be much harder. Acquiring administrator privileges just to push the host client to targeted systems will be a challenge.

It is often necessary to test integration of host monitoring packages across each server or desktop baseline. If constituency systems do not conform to a consistent baseline, testing and reliable operation will be much more difficult. Without careful advanced planning, an enterprise deployment of a HIDS/HIPS, an intellectual property loss prevention tool, or a user activity monitoring system can cause both system and service outages—from applications not starting to printing not working.

- **Multiple monitoring clients deployed on the same host often cause conflicts.**
No one tool does it all, meaning some SOCs leverage multiple independent host-based protecting technologies. In minor cases, one tool will not function correctly because it bumps into the other tool while sinking its teeth into the OS kernel. In extreme cases,

tools will identify each other as viruses and either deactivate one another or cause the system to quit functioning altogether, such as with a Windows blue screen of death (BSOD) [205]. Careful integration testing, along with an open dialogue with involved tool vendors prior to deployment, should help identify and mitigate these conflicts.

- **Host monitoring packages themselves may be avenues of attack or present blind spots in visibility, due to their level of privilege.**
  If a remotely exploitable vulnerability is present in the management interface of a host monitoring tool, it actually introduces a new means for attackers to gain access or spread across the network. Therefore, tool vendors may be subject to additional scrutiny, as their products should enhance, not diminish, host security.

  Malware that exploits other programs with administrator privileges will be on equal footing with the host monitoring tool, possibly undermining or subverting it in a way that cannot be recognized by the SOC analyst. Events indicating normal operation may continue to flow, but the HIPS or AV agent in question may be completely compromised.

  As a result, some host monitoring packages will build their own code base used for direct inspection of system components, memory, and storage instead of, or in comparison to, relying on potentially subverted OS application programing interface for the same functions.

Despite these challenges, virtually all internal distributed or centralized SOCs will pursue instrumentation at the host for a large portion of the constituency. The host offers so many unique opportunities for monitoring and active prevention that such host-based tools are often considered well worth the integration and maintenance challenges.

8.4 Security Information and Event Management

SIEM tools collect, aggregate, filter, store, triage, correlate, and display security-relevant data, both in real time and for historical review and analysis. SIEM workflow is targeted for the SOC, ranging from the ad hoc security team model to a hybrid centralized/distributed model. Best-of-breed SIEM acts as a force multiplier, enabling a modest team of skilled analysts to extract cyber observables from large collections of data. This is a task not easily achievable through other means in a timely, coherent, or sustainable manner. Put another way, the purpose of SIEM is to take a large collection of data and turn it into information, thereby enabling the analyst to turn that information into knowledge that can be acted upon.

By leveraging a robust, scalable architecture and featureset, SIEM can support a number of compelling use cases:

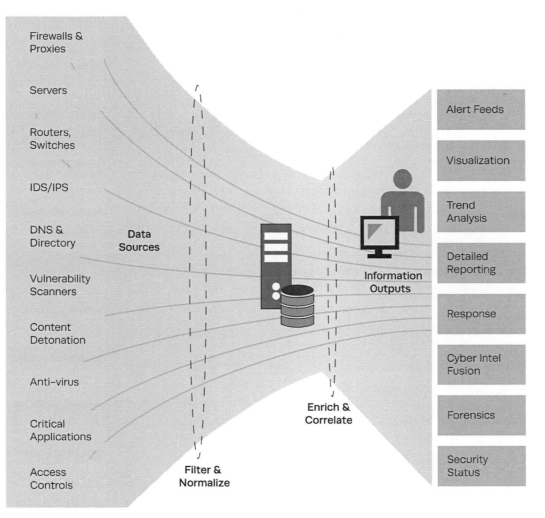

Figure 21. SIEM Overview

- **Perimeter network monitoring.** Classic monitoring of the constituency for malware and external threats
- **Insider threat and audit.** Data collection and correlation that allow for detection and monitoring for profiles of suspicious internal threat activity
- **APT detection.** Piecing together disparate data indicating lateral movement, remote access, command and control, and data exfiltration

- **Configuration monitoring.** Alerting on changes to the configuration of enterprise servers and systems, from password changes to critical Windows registry modifications
- **Workflow and escalation.** Tracking an event and incident from cradle to grave, including ticketing/case management, prioritization, and resolution
- **Incident analysis and network forensics.** Review and retention of historical log data
- **Cyber intel fusion.** Integration of tippers and signatures from cyber intel feeds
- **Trending.** For analysis of long-term patterns and changes in system or network behavior
- **Cyber SA.** Enterprise-wide understanding of threat posture
- **Policy compliance.** Built-in and customizable content that helps with compliance and reporting for laws such as the Federal Information Security Management Act.

SIEM products have been on the market since the very early 2000s. They have proven their value in many industry enterprise SOCs. That said, many organizations struggle to realize the value proposition of SIEM, in large part due to some SIEMs' historical complexity and fragility. Our focus in this section is to understand the capabilities and challenges in leveraging SIEM and to offer some strategies for success.

For more information on SIEM and LM products, the reader may want to consider [206], [207], [208], [209], [210], [211], [212], [213], and the content linked from [214] and [215]. That said, less has been formally written about SIEM than about other CND technologies. As a result, we will dwell on SIEM longer than other tools.

There are a number of vendors who have products in the SIEM and LM market space—too numerous to list here. Instead of providing a comprehensive list, we will refer to the Gartner *Magic Quadrant for SIEM*, which is released annually. The latest, as of this book's publishing, is from 2013, available at [216].

8.4.1 Value Proposition

SIEM can be a big investment—often involving many millions of dollars in software and hardware acquisition, along with the months and years of work required to integrate it into SOC operations. With such a big investment, we should expect a big return. Some SOCs recognize SIEM as little more than an aggregator of massive quantities of log data—this is only the beginning. In order to realize the full potential of SIEM, we must leverage it throughout the event life cycle, as shown in Figure 22.

Figure 22 is essentially a portion of the SOC workflow from Figure 1 in Section 2.2, turned on its side. As we move from the left to the right in this diagram, we narrow millions of events to perhaps a handful of potential cases. In this process, SIEM moves from

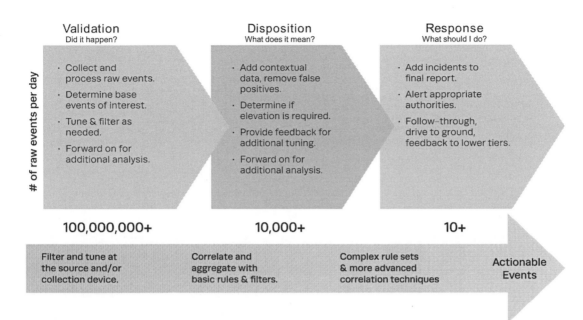

Validation
Did it happen?

- Collect and process raw events.
- Determine base events of interest.
- Tune & filter as needed.
- Forward on for additional analysis.

Disposition
What does it mean?

- Add contextual data, remove false positives.
- Determine if elevation is required.
- Provide feedback for additional tuning.
- Forward on for additional analysis.

Response
What should I do?

- Add incidents to final report.
- Alert appropriate authorities.
- Follow–through, drive to ground, feedback to lower tiers.

\# of raw events per day

100,000,000+   10,000+   10+

Filter and tune at the source and/or collection device.

Correlate and aggregate with basic rules & filters.

Complex rule sets & more advanced correlation techniques

Actionable Events

Figure 22. SIEM: Supporting the Event Life Cycle from Cradle to Grave

automation on the left—through correlation and triage—to workflow support and enabling features such as event drill-down, case management, and event escalation on the right.

Because SIEM acts as a force multiplier, fewer analysts can get more done in a given shift, assuming SIEM has been outfitted with the right data feeds and good content. In fact, one might conclude that a very mature SIEM implementation would actually reduce the number of analysts a SOC needs. This could not be further from the truth, however. Recall our statement from Section 2.7: there is no replacement for the human analyst. In practice, as a SOC implements SIEM, it actually begins to recognize all of the activity in its logs that previously went unnoticed. So, instead of staffing levels going down, they actually go up, because the workload has increased.

The best place for the SOC to be is on the right-hand portion of Figure 23 where a mature SIEM implementation enables a modest team of analysts to achieve what a team of a thousand unaided analysts could not. Even though we may have invested a few FTEs in maintaining and writing content for SIEM, we have more than made up for that with the capability and efficiencies gained.

> **In order for the SOC's SIEM installation to succeed, the SOC must make a sustained staffing investment to leverage it effectively.**

8.4.2 Architecture

While each vendor brings unique features to market, we can identify some common architectural traits:

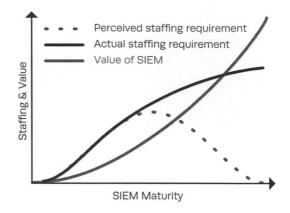

Figure 23. SIEM Value and SOC Staffing Versus Maturity

- A software component processes event data from one or more end devices; this component is often known as an "agent" or "collector," residing in one of two places:
  - On the device, where it has direct access to logs such as through comma-separated value (CSV) files or XML files
  - Remotely, where it either inter-rogates one or more devices for data (pull) or accepts data sent to it (push); the agent can gather this data through various native protocols such syslog, SNMP, Java Database Connectivity (JDBC), or in some cases, proprietary methods such as Microsoft Remote Procedure Call (e.g., in the case of Windows logs).
- Data is collected by the agent, normalized, assigned a relative priority/criticality, and sent to SIEM, usually with controls in place (e.g., mutually authenticated SSL sessions) to ensure successful delivery and to avoid interception or corruption.
- Data is collected at a central location. Data may be stored in a traditional RDBMS, a proprietary backend that supports high-speed queries and condenses on-disk storage, or through a distributed architecture that uses techniques similar to MapReduce [217].
- The SIEM can reduce data volume at several points in its architecture:
  - Aggregation may be applied where multiple events match a given set of criteria and only one is retained. This reduces storage requirements but diminishes available data for use at a later time.
  - Through the use of filters, either at the originating agent or the SIEM collection point, unwanted data can be eliminated due to unnecessary volume, repetition, or lack of value to the analyst.
- The SIEM runs normalized data through a correlation engine in real time using rules targeting various network defense, insider threat, compliance, and other use cases in order to detect complex behaviors or pick out potential incident tip-offs.
  - Some SIEMs also allow the user to run correlation rules against historical data.
  - With proper normalization, prioritization, and categorization, SIEM can fully lever-age various data feeds in a device-agnostic manner.

- Events can have their priority raised or lowered on the basis of hits against correlation rules or comparison against vulnerability scan data.
- Correlation rules can trigger various other user-configurable actions such as creating a case within SIEM and attaching the event to it, running a script, or emailing to an analyst.

- Traditionally, SIEM only brings in event-based data such as NetFlow, IDS alerts, and log data. It can point to other tools by allowing the analyst to fire off an external scriptable command from a right-click in the console, taking the analyst to a third-party tool that gathers vulnerability scan data, malware samples, host telemetry, or PCAP.
- Some SIEMs actually have the ability to automate more complex actions resulting from correlation rules (e.g., deactivating a VPN connection or changing a firewall rule). Use of such features often has the same implications as putting an IPS into active prevention mode: poorly tuned correlation rules can lead to a DoS.
- Raw and/or normalized data is stored to provide an audit trail and a knowledge base for investigations, pattern analyses, and trend reporting.
- Analysts interact with the data through various output channels, either through real-time scrolling alerts, event visualizations (bar chart, pie chart, tree maps, event graphs, etc.), or through static means such as ad hoc or scheduled reporting.
- The system provides some level of incident ticketing or case tracking, allowing users to acknowledge and escalate both alerts and cases.
- SIEM users can view content created by other SIEM users whose access is controlled through groups and permissions, leveraging both vendor-supplied "stock" content as well as customized reports, rules, filters, dashboards, and other content.
- Some best-of-breed SIEMs provide methods for users to move SIEM content in and out of the SIEM. This functionality may be further enhanced through online SIEM user communities where users can share and collaborate on content.
- Best-of-breed SIEMs support multitiered, peered/clustered, or redundant deployment scenarios:
  - With tiering, one SIEM can forward some or all of its alert data to a parent SIEM, leveraging the same agent or collector framework that gathers data from end hosts.
  - With peering or clustering, each SIEM collects a different set of data, and when the user runs a query, the work is spread across multiple SIEMs, speeding query times.
  - In redundant scenarios, multiple SIEM instances can ingest the same data, potentially through "dual reporting" from one agent to multiple SIEMs, or through syncronization between disparate SIEM nodes.

- In any case, we can scale enterprise deployments beyond hundreds of millions of events per day while still providing meaningful data to the analyst, with reasonable query times.

Figure 24 depicts several delivery options for data, along with load-balanced or redundant instances of a SIEM or LM appliance.

In this example, NIDS, firewall data, and workstation data are being collected through protocols native to each device—JDBC, syslog, or Windows Remote Procedure Call. The agent can sit either on the system generating the data (as is shown with the domain controller) or, perhaps, remotely (as is shown with firewall data and IDS events). The point here is that a best-of-breed SIEM tool should provide multiple options for data delivery and collection, along with redundancy and failover.

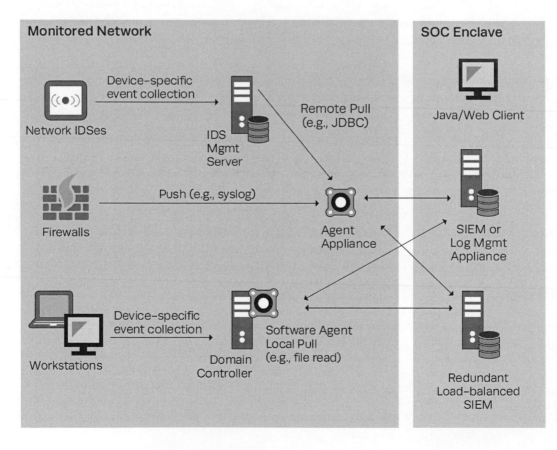

Figure 24. Log Data Delivery Options and SIEM Tiering

8.4.3 Log Management

Collecting and querying events from a disparate set of systems or applications doesn't always necessitate the features and cost associated with a full-blown SIEM. Oftentimes an LM system, which is usually simpler to set up and use, is a better choice. LM systems incorporate the aggregation, storage, and reporting capabilities found in SIEM, but with significantly streamlined interfaces, simplistic analytics, and little correlation. Splunk (the company) probably puts it best when it describes its LM capability as "Australian for grep" (a takeoff on the old Foster's beer commercial), suggesting that Splunk (the product) is a supersized, more capable version of the familiar UNIX text search tool [218]. That said, many vendors describe their LM systems as having features found in a full-blown SIEM. Therefore, understanding where a given product falls in the SIEM and LM spectrum can sometimes be challenging.

The SIEM architecture described above also describes the components of most LM solutions; however, we note the following differences:

- LM systems usually lack a robust correlation engine, advanced long-term analytics, full-fledged workflow, and escalation support.
- LM systems typically perform less preprocessing and normalization on data feeds, focusing instead on very fast ad hoc search capabilities. Robust correlation usually requires a SIEM to understand the meaning of different data elements in each event, because most LM systems don't fully parse their data. Instead, they focus on quick, full-text search or specific-only feature extraction—truly robust correlation either isn't possible, or requires much more work on the content author's part.
- LM systems, because they do not perform nearly as much preprocessing or postprocessing of data, can ingest it substantially faster, with a smaller code base. This is a very intentional design trade-off and, as we can see, has its pluses and minuses.

In many regards, the two products are seen as complementary—most SOCs that implement SIEM started with basic LM and grew their data ingest and query capabilities as their mission expanded.

Many smaller SOCs, which don't have millions of dollars to spend on SIEM, may choose to augment their native IDS consoles with an LM appliance, thereby getting some of the benefits of SIEM but at a fraction of the cost. In addition, it is common to see IT shops or NOCs deploy LM systems on their own, without a specific interest in CND. Finally, security organizations may choose to deploy LM systems for their own audit purposes. In either case, it helps if the SOC can pool resources with these groups to unify their data collection architectures. We mentioned this in several SOC capabilities in Section 2.4.

In order to better summarize the difference between a full-blown SIEM and log aggregation devices, let's look at some key qualities of both. While a given SIEM or LM product will diverge from Table 12 in one or two ways, we can make some generalizations:

Table 12. SIEM and Log Aggregation Systems Compared

What	SIEM	LM
Data types	Broad selection	Broad but sometimes more limited
Data normalization	Usually robust	Usually simplistic
Datastore	Commercial database or custom	Usually custom
Correlation	Robust	Simple/none
Real-time alerting capabilities	Excellent	Fair to none
Historical trending capabilities	Good to excellent	Fair to good
Ad hoc data query	Fair to good	Excellent
Reports	Good to excellent	Good to excellent
Event ingestion rate	Fair to good	Good to excellent
Structured query speeds	Fair to good	Good to excellent
Unstructured (full-text) query speeds	None to poor	Good to excellent
Client interface	Complex/full-featured, Java, or Web 2.0-based	Simple to complex Web 2.0-based
Form factor	Software, virtual appliance, hardware appliance	Software, virtual appliance, hardware appliance
Typical cost (per instance)	$50,000–$1,000,000+	$0–$100,000[1]

1 In an effort to increase product exposure and competition, some LM vendors offer free or "nearly free" scaled-down software versions of their product. These offerings are, of course, not intended for major deployments as they usually have enterprise-focused features removed or are limited in terms of event volume, event retention, or both.

While there is some overlap among network management, SIEM, and LM, SIEM covers several areas of functionality key to CND that LM systems do not. (See Figure 25.)

One way to look at the difference between full-blown SIEM and LM systems is that SIEM can serve as the cornerstone of CND workflow, whereas a LM system cannot. Recall our discussion of SOC organizational models from Section 2.3.2—there are many constituencies where CND is performed in an ad hoc manner (e.g., with a security team). These organizations have few resources and do not devote many (if any) full-time staff to CND. Thus, their needs are well satisfied by an LM appliance. Full-fledged SIEM requires care and feeding that only medium to large SOCs can provide.

8.4.4 Acquisition

SIEM is probably the largest single purchase a SOC will make. Given its high cost, we will discuss SIEM tool acquisition. Before considering purchasing a SIEM tool, the SOC should consider the following baseline conditions:

- The SOC has log aggregation tools already in place; its needs with respect to analytics and correlation exceed those that its current tools offer.
- The SOC performs a substantial portion of its analysis duties on real-time data.
- The SOC has identified multiple data feeds beyond IDS/IPS that it intends to feed to a SIEM in a sustained, real-time fashion.
- The SOC engages in a sustained sensor management and tuning process with dedicated staff, thereby suggesting it is ready to take on SIEM management.

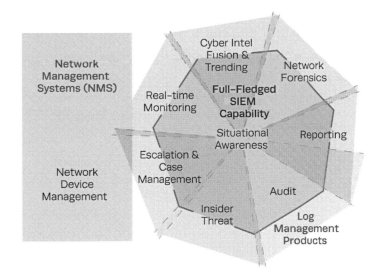

Figure 25. Overlap Between SIEM, Network Management System, and LM

There are certainly exceptions to each of these conditions, but these hold true in the vast majority of cases where a SOC is contemplating SIEM acquisition. For further details on this, see page 11 of [219].

The next logical consideration in acquisition is requirements. Requirements are what drive acquisitions—business, functional, performance, system, user, operational, and the like. What does the SOC want to get out of the capability? What major use cases will it serve? Does its architecture support where the SOC will be in three or five years? These are just the beginning.

As with any major acquisition, the SOC may want to consider a bake-off of two or three major SIEM vendors with an on-site pilot. In this case, all tools can be compared side by side against well-defined, repeatable, measurable requirements. SOC engineers may wish to leverage a scored, weighted, repeatable requirements comparison chart such as Kepner-Tregoe [220], especially if they have a large number of requirements or face a large acquisition. If vendors are working with a potential customer who is serious about the acquisition objectives, they will often be willing to grant a 60- or 90-day temporary license for such purposes. Even if it's a pilot of just one product, this can be really helpful in driving requirements and managing expectations.

SIEMs can sometimes have fairly complex licensing schemes. Each vendor will likely base its product cost on one or more of the following:

- Number of "master nodes" (e.g., managers that hold the "brains" of the SIEM, such as the correlation engine); sometimes measured in the number of CPU cores belonging to one or more master nodes
- The amount of data ingested by the system, often measured in gigabyes per day, events per day, or sustained events per second
- Number of users accessing the system (e.g., the number of "seats")
- Number of appliances purchased (in the case of virtual software or hardware appliances)
- Number of devices sending data to the system and, possibly, the number of device types (Windows, UNIX syslog, database, application, etc.)
- Additional features or add-ons such as content packs.

In some cases, SIEM vendors will actually place hard limits in their devices on the basis of the license negotiated when the product was acquired. Operationally, this can be frustrating when a SOC hits a hard limit on event ingest rate, the number of devices it's collecting from, or the amount of data retained by the system. Sometimes these are hard predictions to make when first buying a SIEM, so careful planning is key. The SOC should also plan for out-year costs as part of its TCO. Annual maintenance and support fees frequently measure between 20 and 30 percent of the initial acquisition cost.

8.4.5 Calibrating Expectations

When getting started with a best-of-breed SIEM, it is sometimes challenging to know what to expect. In Table 13, we examine several key parameters of SIEM architecture and implementation, giving some *order of magnitude* estimates of what is currently possible with commercial offerings.

Table 13. Best-of-Breed SIEM Characteristics

Characteristic	Discussion
Deployment package—agents	Agents (also referred to as collectors or connectors) are usually available as software packages that can be installed either on the end device at a natural collection point (such as an IDS database or syslog server) or somewhere else (such as a freestanding agent system). While there is no such thing as a truly "agentless" architecture, best-of-breed SIEM will be respectful of resource demands placed on the hosts that originally generate or transmit event data.
	Software agents should have several configuration options, allowing administrators to optimize their CPU, memory, disk footprint, and multithreading. Administrators should be able also to optimize the load agents place on systems they pull data from (especially databases) (e.g., through how often they query for new events and how many events can be brought back at one time).
	Some vendors will also sell agent hardware agent/collector appliances, which can be easier to manage. If this is the only choice, deployment options may be limited and the SOC may be stuck paying for more appliances than it needs.
	Regardless of specific vendor implementation, agents should provide robust support for current versions of popular data sources and interpret that data in its original (often proprietary) format, preserving the richness and meaning of data as it was first generated.
Deployment package—master node/manager	Many SIEM vendors ship both a hardware appliance version of their product and a software version—either as a traditional piece of installable software or as a virtual appliance. Echoing trade-offs from NIDS in Section 8.2, there are pluses and minuses to either approach. In order to scale SIEM to really large installs, deploying as a software package on the SOC's choice of enterprise-grade server hardware (usually Linux on x86-64) is almost a necessity.
	LM products may also come as software packages or virtual appliances, but vendors lean more toward hardware appliances since they fit into the themes of easy deployment, management, and support. This is changing somewhat as SOCs and large IT enterprises consider LM in a purely "cloud" architecture.
	Some SIEMs will split the master node function between two systems—one that does the correlation and another that hosts the datastore, possibly in an RDBMS. Others may balance event storage and query across one or more nodes running a distributed "NoSQL" backend. Simpler implementations (e.g., those using a proprietary datastore) and virtually all LM products will couple the front-end manager together with the backend datastore on one physical system.
	The master node/manager should be properly multithreaded in order to support parallel query execution and efficient correlation.

Table 13. Best-of-Breed SIEM Characteristics

Characteristic	Discussion
Data retention	Both SIEM and LM products are theoretically capable of storing many years or, perhaps, even decades of events. What usually prevents SOCs from retaining data this long is how long they've had the product in operation since its first iteration. Some SIEM and LM products don't make data migration from one major upgrade to the next as easy as it really should be. In some cases, a SOC may choose to implement a new instance of a SIEM in an effort to start with a clean slate. This will become a greater issue as SOCs cycle through many generations of SIEM products, while increasingly looked to as the shop for long-term audit data retention.
	As products have matured and stabilized, the upgrade path for product software has provided better support for retention of old event data, along with old custom content.
Datastore size	At the time of this book's writing, mature SIEM installations typically have datastores of from a few to perhaps 50 TB or more of storage. Going beyond the 10 or 20 TB mark challenges the storage capabilities of some products. Many vendors that incorporate proprietary backends advertise a 5:1 or 10:1 data compression ratio compared to a relational database. Given loads of 100 million events per day, it is not unusual for an RDBMS to retain 60–90 days of data, whereas a proprietary compressed datastore of the same size may retain 6-12 months of events or more.
Number of events	A healthy SIEM implementation usually measures event ingest rates in the range of thousands of events per second, which equates to tens or hundreds of millions of events per day. As shown in Figure 9, above, this is the sweet spot for SIEM and LM. Although many SIEM vendors advertise ingest rates of 10,000 to 100,000 or more events per second, we must ask, What is the use case for bringing this much data into one place? Would the analysts be better served by breaking this up into tiers, and/or are we being too indiscriminant about what data we're collecting? It is not unusual to see a SIEM or LM appliance hold many billions of events online; getting past the trillion event mark is rarer, however.
Number of users	Most SIEM and LM vendors have built their products to deal with the typical pool of analysts in a centralized or distributed SOC, which usually comes in around 10–50 people *actually logged into and using the product at a time*. Large SOCs that incorporate elements of the distributed organizational model such as "forward deployed" analysts may run into scalability challenges, especially if those extra users have custom content and reports that further obligate system resources.
	It's also worthwhile to note that geographic distribution of users shouldn't have a major impact on console usability, provided reasonable bandwidth is available (i.e., they are not accessing the SIEM through a dial-up connection or a heavily saturated link). When a SOC wishes to support more than 40–50 concurrent users of a SIEM, it may look at a peered, clustered, tiered, or some other distributed approach to spread for different classes of use cases. For instance, a SOC may give power users their own SIEM instance, so expensive queries do not dominate SIEM system resources. This, then, provides an acceptable user experience for all members of the SOC.

Table 13. Best-of-Breed SIEM Characteristics

Characteristic	Discussion
Number of devices	In reality, this should not be a limiting factor (beyond any artificial limits due to the licensing models) either at the collector or master node. A SIEM should be able to handle tens or even hundreds of thousands of end devices across its various data feeds (which is not uncommon when instrumenting a large enterprise). The SIEM should be able also to recognize multiple data feeds from the same end point (e.g., AV, Windows security event logs, and vulnerability scans). Well-architected agent software should be able to cope with many devices (perhaps tens of thousands) whose aggregate event flow may sustain more than a thousand events per second.
Query response	Most questions an analyst will ask of a SIEM or LM appliance are expected to come back in a matter of seconds or minutes. Given our typical scenarios where we collect many millions of events per day, this may mean well-formed queries can go back several days with RDBMS backends, or perhaps several weeks with proprietary backends. If, for instance, we search for an IP over a day of data and don't get the answer back in a minute or two, we may want to examine our data ingest policies or datastore optimizations.
	More complex questions, like long-term trending, anomaly detection, and unoptimized queries that trigger the equivalent of a full table scan, may take a lot longer to execute. In these cases, content must be carefully scheduled so as to not dominate system resources, especially during core business hours.
	Most historical assumptions about SIEM and LM performance are based on spinning hard disk backends and RDBMS-style datastores. With solid state storage and MapReduce query engines, new opportunities for asking much more complex questions of data, over longer periods of time, are possible. SOCs looking for advanced long-term analytics or query capability to support large user loads may (1) consider SIEM backends that leverage some or all of their event storage on SSD, or (2) leverage cloud technologies that leverage MapReduce techniques for scaling.
	Most SIEM vendors that provide hard numbers about system performance express them in event ingest rates. As we've discussed, query performance is often much more of a factor and can be at odds with ingest rate. Ultimately, the SOC is best able to make an informed purchasing decision with regard to query performance if it is able to work with the product hands-on, most favorably in a preacquisition on-site pilot with live data and real analysis.

Again, these metrics are provided as a rule of thumb to help SIEM architects and maintainers gain some context for what is reasonable to expect of their implementations. As this book was written in 2014, the reader may want to factor in general growth for IT systems in subsequent years.

8.4.6 Observations and Tips for Success

As with any other tool discussed in Section 8, an entire book could probably be written on SIEM. Our emphasis here is on sound decision making from an architectural and operational perspective. As a result, we will round off our conversation with some lessons learned. These takeaways are written as they apply to SIEM but also have strong bearing on LM tools.

- **Security and network management tools are not interchangeable.**

  SIEM and network management systems (NMSes) have many similar architectural features, such as the ability to aggregate lots of log and alert data into one place, process it, and visualize it. The confidence we can place in logs and alerts for security products tends to be less than in other areas of IT such as networking. When a router says its fans are spinning slowly and a switch says a port is down, that is almost certainly the case. When an IDS says, "You've been hacked," chances are everything is okay. SIEM has a rich feature set that supports the workflow necessary to drive events to ground, evaluating whether a given alarm is a true or false positive.

  NMSes lack many of these security-specific features; enterprises seeking to maximize value for network and security management are, therefore, advised against trying to combine network management and CND workflows into one system.

- **The best SIEMs were built from the ground up as SIEMs.**

  Many LM and SIEM products in today's marketplace were first architected and coded with a very narrow scope of features and use cases, compared to what they currently advertise. Just as it is important to lay a strong foundation for a tall skyscraper, a good SIEM product is built with a scalable backend, robust correlation engine, and extensible data-ingest capabilities.

  Some SIEM products lack a strong foundation and, as a result, have run into problems as developers bolt on more and more features. A poor foundational architecture can manifest itself through poor ingest rates, slow query speed, fragmented workflow, lack of key use cases, poor or no real-time visualization, lackluster user interface capabilities, and clunky reporting.

  When a SOC is acquiring a SIEM, it is important to look for features that suggest the product was built as an enterprise-grade SIEM from the start, not as a one-off or homegrown project that later turned into a commercial offering.

- **Consider the whole package.**

  When contemplating an LM or SIEM purchase, many SOCs are narrowly focused on one criterion—"Can the product ingest data type X?" This is probably most analogous to buying a car solely based on what tires it comes with or what color paint it has. These are certainly important features, but (1) they can be changed by the owner at modest expense, and (2) there are vastly more important considerations such as speed, reliability, and operator experience.

Most good SIEMs can accept almost any possible security-relevant data feed, either out of the box or through an agent software development kit (SDK) of some sort. What's not important is, "Can the SIEM ingest my data?" That's a given; what's important is, "What can I do with the data once my SIEM has vacuumed it up?" This means everything from real-time dashboards to correlation to reporting and escalation.

- **You get what you pay for.**

SIEMs are often very expensive, but there's a reason for this; building an enterprise-grade product that can ingest and store billions of events from thousands of devices, support dozens of concurrent user queries, and streamline their workflow and interaction cannot be done in a weekend. Many SIEM products were originally built by small startups that sell to medium and large businesses and government organizations, of which there are a finite number.

Many LM products exist to fill the lower cost market space and therefore should not be considered as true SIEMs. In many cases, a SIEM with a narrow feature set or a simple LM appliance is the best choice, not just from a cost perspective but also in terms of learning curve. SOCs that have a smaller constituency and no plans to integrate a large SIEM capability may be best served by focusing on lower cost LMs.

Conversely, SOCs that start with a low-cost SIEM face the biggest financial and political hurdle when the product doesn't live up to original expectations or growing mission demands, forcing the SOC to jettison the initial product and time investment in favor of a more robust, more expensive offering.

Whereas SOCs' network IDS or vulnerability scanning needs can be met with world-class FOSS offerings (Snort and Nessus, respectively), fewer options exist for FOSS SIEM. While some exist (e.g., Open Source Security Information Management [OSSIM] [221]), consider the complexity in building and maintaining a robust SIEM— not to mention a growing portfolio of data types and agents. Furthermore, most organizations in search of a true enterprise-grade CND data aggregation and workflow tool have deep pockets. In the case of Snort, a very large user base is able to drive continual improvement to a complex codebase. SIEM, by comparison, has equal or greater complexity but fewer likely customers. That said, there are some FOSS tools (e.g., S/GUIL) that succeed at supporting real-time CND analysis with multiple data flows, in part because they don't attempt to take on the entire SIEM feature set. FOSS tools that focus on LM without the complexity of full-fledged SIEM also have the opportunity to offer compelling features and performance (e.g., Logstash [222] and Kibana [223]).

- **A day to install; a year to operationalize.**

The initial setup of SIEM is largely straightforward and can usually be accomplished in a day or two for a single enterprise-grade instance. In a few weeks or months, the first few data feeds can be hooked up and tuned, with content created that provides quick wins. However, outfitting the system with the right data, tuning it, writing content for the constituency, and integrating it into operations altogether takes a year or more—*for political and process reasons, not technical reasons*. In many cases, getting data owners to provide reliable, sustained audit log feeds can be a politically arduous process. As a rule of thumb, the more robust the SIEM, the steeper the learning curve—not because the interface may be clunky (although this is sometimes the case), but because it takes even the smartest analysts time to understand the fundamentals of the tool and what it is capable of. A college CS student can learn the fundamentals of C or Java in several weeks; becoming truly proficient in a language takes much longer. Essentially, the same is true of SIEM. One must also consider the time commitment for integrating the features of SIEM into SOC operations (e.g., user training and writing SOPs).

Many successful SIEM implementations worked because they had a champion within the SOC who understood the technology, invested the time necessary to get data feeds working, and adapted content to the constituency environment. Consider this—for every neat whiz-bang use case shown by the vendor in the product demo, it is likely that some administrator or analyst will have to write or tweak that feature before use. *This is the case for every SIEM product.* Out-of-the-box content serves as a good start for most SOCs, but the best content is often written from scratch.

- **Each part of the SOC will use SIEM differently.**

Each work center in the SOC has different uses for the data and features SIEM can provide. Tier 1 will be interested in real-time triage of data and concrete use cases they can write SOPs for and put in front of junior analysts. Tier 2 will likely be interested in gathering as many details about a potential incident as possible, meaning they will run free-form queries over long periods of time. Those responsible for attack sensing and warning or trending activities will likely fuse various sources of cyber intel and tippers into the tool, repeatedly running long, computationally expensive queries. Managers will be interested in case management and metrics from the tool, validating that their respective part(s) of the SOC are following procedures, following up on anomalous activity when the system catches it, and not letting cases languish in the system.

Each of these parties has an equally valid need for training on the tool, well-written content to fulfill their mission role, and performance that meets their ops tempo. Many SOCs meet these needs by designating a SIEM "champion" as a "content manager" similar to that of an IDS tuner. Moreover, each SOC section will have overlapping needs for the tool, and it is important that one person ensures the content and that queries are effectively consolidated and deduplicated. While regular maintenance such as database tuning, agent deployments, patching, upgrades, and the like are all important, having dedicated staffing for SIEM content development and management is key to SIEM success.

- **Many parties outside the SOC—security officers, sysadmins, and CISOs—can directly access raw, security-relevant data with SIEM. Should they?**
  There are many stakeholders in the constituency who have a legitimate need to work with security logs on a regular basis (e.g., security has to perform system auditing). SIEMs can really help with this. However, before a SOC invites everyone in, it needs to consider three issues:

  First, does the SOC want someone outside the SOC finding an intrusion before it does? More than a few CISOs have expressed a desire to have a real-time view into the security status of their enterprise. Consider, however, if the CISO sees something in the console and picks up the phone before the SOC has all the answers or has even noticed what the CISO is looking at—not a good position to be in. In many cases, the CISO or CIO is just as happy getting weekly or month roll-ups of significant events or weekly metrics. Or, at the very least, the SOC can be trusted to call when something bad is actually happening.

  Second, not everyone needs access to all the data. If the SOC does open up access to external parties, it is very likely they only need to see a slice of data that applies to their portion of the enterprise. Perhaps they don't get to see events in real time, and they can query events older than 24 hours. The SOC's mission is, therefore, not diminished—it is still the go-to shop for real-time cyber SA.

  Third, and perhaps most important, providing large swaths of security logs to a variety of parties will eventually compromise insider threat cases. AV should alert when users download well-known password-cracking and port-scanning software. FTP logs can reveal information leakage and exfiltration of data. Application and Web server logs may indicate malicious or inappropriate behavior on the part of privileged users. Cases involving any of these activities could be blown if word gets back to the perpetrator before the authorities have completed their investigation. SOCs are well advised to consider what raw data is shared with whom. Alternatively, the risk can be further minimized by providing scheduled roll-up or summary reports instead.

- **New tools mean new processes.**

Even the most basic SIEMs and LMs enable analysts to triage and analyze event data more efficiently than without such a tool. Despite this, when SIEMs are introduced to a SOC, many analysts stick to their old ways of looking at data. Using a SIEM to triage and view data in the same way as the native console provides little added value. When a SIEM is brought in, analysis SOPs need to be modified and training given to break users of their old habits and better utilize their new tools.

This may be as simple as tailoring views for Tier 1 such that they're looking at only the most important events (which constitute .001 percent of all the data collected). These events may be still simple, real-time scrolling events, but at least the system is providing value by automating a portion of data triage. Analysts must be freed to analyze; enforcing monotony will prevent team members from exercising their curiosity and, most likely, drive away those with the most talent.

One of the big selling points for SIEM is the ability to bring disparate data feeds into one console. It should be noted, however, that there is nothing wrong with having multiple different views into the data. Some Tier 1 ops floors will separate out different streams of triage data into various dashboards which can be split up (or shared) amongst different users. This value proposition works only if all event-type security data feeds consumed by the SOC are directed at, and triaged by, the SIEM. Pointing only some feeds at the SIEM while neglecting others diminishes this value.

- **A SIEM is only as good as the data you feed it.**

The old saying, "garbage in, garbage out," applies perfectly to SIEMs. We have discussed how the value of even the most relevant, detailed security logs can be completely diluted if we're not discriminating about everything we bring in. It is of utmost importance to select and tune data feeds according to the constituency environment, threats, vulnerabilities, and mission. (See Section 9.2.)

One of the by-products of a healthy selection of data sources is that a SOC's IDSes are essentially put on the same footing as any other source of data (e.g., Web proxy records or application logs). From the perspective of the analysts, IDS alerts are just another data feed among many feeds they can choose from when tailoring a report, correlation rule, or dashboard to a given threat.

- **Lack of a single common data standard can be overcome.**

The history of audit data aggregation and security data management is paved with industry standards, none of which have had comprehensive adoption: Common

Intrusion Detection Framework (CIDF) [224], Incident Object Description and Exchange Format (IODEF) [225], Security Device Event Exchange (SDEE) [226], WebTrends Enhanced Log File (WELF) [227], Common Event Infrastructure/Common Base Event (CEI/CBE) [228], Common Event Format (CEF) [229], and Common Event Expression (CEE) [230]. A common theme among them all is the desire to provide a vendor-agnostic format for recording and ingesting security-relevant data.

Inspection of leading SIEM products will yield dozens, if not hundreds, of data parsers, each for a different device type and vendor. SIEM vendors expend a lot of resources to keep these translators up to date. This can be especially frustrating for SOCs with less popular or custom data sources they wish to integrate in the SIEM, even with a good parser SDK from the vendor.

As of this book's writing, convergence on a standard has not occurred, even though we have several. As a result, SIEM vendors and most of their customers continue with the status quo. Most organizations that ingest more than a handful of data types cope with this situation fairly well (although it does consume some resources in updating agent parsers and running down bugs in missing or garbled data). That said, there are a few acute pain points. The first is with moving data in and out of a SIEM, especially when consuming the data from another SIEM. The second, as we mentioned above, is with any SOC that needs to consume data from multiple custom applications that most likely aren't supported out of the box by a given vendor. Third, many LM products choose to ingest with little post processing, leaving the data in a relatively raw form.

- **Automated response capabilities present the same challenges as IPS.**
  Some SIEM vendors have advertised features in their product whereby a correlation rule can trigger an automated action. Sometimes known as "automated course of action" tools, the SIEM can trigger actions like a firewall rule change, user account deactivation, VPN session termination, or anything else that can be scripted. As we discussed with NIPSes, great care must be taken when implementing any kind of automated prevention. With SIEM, we have an even greater chance for false positives and glitches in end-device integration.

  This feature is best used in a high-visibility line of business where there is a large pool of privileged users who are under increased scrutiny—perhaps because their actions present financial or mission risk to the enterprise. Financial accounting or IT call centers are excellent examples. Here, user actions can be closely monitored, and there are well-defined business rules that define suspicious or disallowed actions.

Very few SOCs have reached a level of maturity where automated response actions might become a realistic option. Before exercising this capability, the SOC may consider more complex correlation use cases such as using watch lists or having automated response actions prefetch information (e.g., PCAP data or vulnerability scans) to the analyst.

- **SOCs should architect their collection and retention to support criminal, civil, and administrative proceedings.**

  Ultimately, the electronic evidence a SOC collects may be used by law enforcement, legal counsel, and various investigative bodies, in response to serious incidents. Just as with any artifact collection procedure, the SOC should ensure that the way it gathers, stores, and analyzes security-relevant data supports these activities. Moreover, applicable privacy laws may restrict the SOC's ability to collect or retain certain log types of content. While computer forensics is out of the scope of this book, the SOC may wish to discuss this with legal counsel and examine applicable laws and regulations for further guidance on this matter. (See [231] and [232].) Also, it should be noted that the interpretation of such laws varies widely and can have a profound impact on the cost of the SOC's log collection and storage architecture. Ensuring that common sense is integrated into system design, along with a well-informed understanding of the law's impact, is critical.

Chapter 9
STRATEGY 7: Exercise Discrimination in the Data You Gather

SOCs wishing to gain visibility into their constituencies pursue two complementary strategies: (1) they leverage security-relevant data feeds, and (2) they deploy their own network and host sensors. Acquiring and deploying these capabilities soaks up tremendous amounts of resources, in large part due to the scale and complexity of deployment and the cost of collecting, retaining, and processing the data generated. Too little data means we can't find or follow up on intrusions. Too much data means the good stuff gets lost in the noise. We wish to address two common problems for SOCs: (1) critical blind spots in coverage, and (2) data flows not being tuned properly.

In our seventh strategy, we examine approaches to maximizing our resources when it comes to instrumenting the enterprise. Our goal is to gather the right data in the right amounts from the right places in the enterprise, with the minimal amount of effort and expense.

9.1 Sensor Placement

In this section, we discuss where we should place sensor technologies, including:

- Passive network sensors, including general-purpose NIDS
- Active network sensors: NIPS and content detonation devices
- General-purpose HIDS/HIPS
- Other host-based instrumentation and protection such as configuration monitoring and remote forensics support
- Application-specific protection appliances (XML, database, etc.).

The choice of technologies used and where to place them is squarely within the CND program, making it the SOC's job. Participation from other parties such as the CISO, security, and IT operations is often called for, but preferably the SOC has the lead role. There are many drivers to how a SOC instruments its networks. Principally, we are looking at maximizing coverage throughout the cyber attack life cycle. SOC resources are finite, so we must make careful choices in terms of what technologies to acquire and where to employ them.

While we discuss a range of technologies, there are some common themes to effective placement:

- Match the monitoring technology to the current threat landscape and asset type.
- Provide maximum breadth and depth, given a finite number of sensors.
- Guard connections between enclaves and differing trust levels.
- Provide the CND analyst with relevant, timely, and rich details on network and host activity of concern, covering the entire attack life cycle.
- Balance resources among competing priorities, such as asset connection to core constituency mission and their exposed attack surface and vulnerabilities.
- Plan for TCO, including acquisition, operation, and maintenance.

Many SOCs tend to focus their monitoring resources on the "front door" to their network, such as their Internet gateway(s). While this is usually the top priority, it's really just the beginning. For instance, a constituency's core mission systems may communicate with the Internet through separate connections from general email and Web traffic. There is a natural tendency to put a network sensor wherever there is a firewall, yet "forgotten" backdoor connections must not be left out since they often pose an appealing infiltration or exfiltration point for the APT. And that's just the perimeter; we will also look at how to instrument internal segments of the constituency. By placing monitoring where the adversary is most likely to be but does not expect to be seen, the SOC stands a better chance of spotting intrusions sooner.

9.1.1 Passive Network Sensors

Passive network sensors—signature-based IDSes, NetFlow sensors, and the like—almost always comprise a plurality of the SOC's monitoring footprint across its constituency (measured in terms of SOC analysts' attention). Furthermore, these technologies form the backbone of the SOC's SA and are often the first monitoring capability a SOC will deploy.

When we discuss architecture and logical placement, we are considering the complete package of technologies discussed in Section 8.2, along with Section 8.1.4:

- Signature and/or heuristics-based IDS monitoring
- NetFlow and/or traffic metadata "superflow" record generation

- Sustained and ad hoc full PCAP collection
- Passive OS and service fingerprinting.

Referring back to Section 8.2.9, *even if a SOC completely eschews classic signature-based NIDS, it will need some sort of enterprise-class passive network monitoring package that can be deployed widely.* When we prioritize passive network sensor placement, we are looking to meet several competing goals with respect to a potential sensor PoP. They are shown in Table 14.

Constituency Internet gateways are considered the most obvious and immediate choice for IDS placement. This sensor placement at these locations meets most of the goals we discussed: (1) mission-critical systems usually connect through it, (2) a large proportion of the entire constituency's traffic goes across it, (3) systems on the other side (the Internet) are completely untrusted, (4) we usually see lots of unencrypted traffic going across it, and (5) we can expect that many constituent systems will expose various vulnerable services through it. Many enterprises that don't even have a SOC will often choose to place a NIDS or NIPS at their Internet gateway, with monitoring duties falling upon their general IT department or NOC.

Beyond major enterprise perimeter points, the choice of network sensor deployment becomes more complicated. We must follow the law of diminishing returns, considering our limited resources. Let's look at some additional considerations for network sensor placement:

- **Most monitoring products target TCP/IP and Ethernet.**
  Individuals with a background in WAN technologies who approach the topic of network intrusion monitoring will often look at WAN gateways as logical points for IDS placement. This presents two problems: (1) most intrusion monitoring technologies operate on Ethernet links and don't know how to decode an asynchronous transfer mode (ATM) cell (for instance), and (2) these kinds of links often involve high bandwidth (e.g., 10 Gb and above).

  It is less common for SOC personnel to have an in-depth knowledge of WAN technologies such as multiprotocol label switching (MPLS), synchronous optical network (SONET), and ATM, and, therefore, the ability to understand the content or placement of a sensor if it were placed at this layer of an enterprise network.

  As of this book's writing, most network monitor technologies operate comfortably in the 10 Gb and below range, with some technologies extending to 40 Gb. When we consider full-session network capture, the situation is further complicated.

- **Projected changes in the enterprise architecture**
  Will the PoP have a large number of assets deployed on either side of it in the near future? Conversely, is the PoP about to be decommissioned?

Table 14. Network Sensor Placement Considerations

Goals	PoP Example(s)
Gain visibility into systems important to constituency mission	Servers hosting custom mission applications and sensitive data sit behind PoP.
Provide coverage for systems that are of especially high value to adversaries	Systems behind PoP contain trade secrets, source code, or confidential records. An Internet-facing email gateway serving a large user population
Achieve greatest "bang for the buck" by picking PoPs that host a large number of network connections (e.g., network "choke points")	All network traffic between two major corporate regions transit PoP, covering 10,000 systems
Protect systems that sit on the trusted side of a controlled interface (e.g., a firewall)	PoP is between university dorm networks and the university's registrar's office. Company A's servers communicate with Company B's servers across a private link.
Have complete insight into the traffic being observed (e.g., it is not encrypted and uses protocols the sensor understands)	On the unencrypted side of a VPN termination point or SSL accelerator On both sides of a NATing firewall or Web proxy[1]
Leverage passive monitoring as a compensating control for systems that lack critical security features or have serious unmitigated vulnerabilities	Unpatched systems providing various services to systems on the other side of the PoP, with no firewall protecting them Legacy systems that cannot be patched and respond nondeterministically to incorrectly formatted protocols

1 Networking proxy devices, such as Web content filters and NATing firewalls are an interesting (if not maddening) case for network sensor placement. Tap points are usually needed on either side of the device in order to see the true source and destination hosts involved. Outside the proxy, all we see is traffic from the proxy to various Web servers. Inside the proxy, we see corporate users surfing the Internet, but we can't be certain of the IP of the Web server providing the data. Even with sensors on both side of the proxy and proxy logs, this can be a challenge due to how some devices translate traffic and cache data. Engineers must carefully determine the right mix of sensor technologies and logs that will give analysts clear visibility into traffic transiting the proxy.

What kind of bandwidth will be seen at the PoP in the near future, and, therefore, how much overhead should be left in the chosen hardware and software?

One key success point is to work with network owners to acquire throughput statistics for proposed sensor PoPs prior to hardware acquisition and deployment. This allows SOC engineers to plan for hardware resources needed, taking into account protocols that can be filtered out of the monitored stream.

- **Maintainability and access**

   Is the PoP physically located where sensor(s) have good connectivity and bandwidth back to core SOC systems?

   Can SOC sysadmins physically access the equipment, or can a TA at the site perform touch maintenance when needed?

- **Existence of other monitoring capabilities**

   If the SOC can't put a sensor at a given location (for whatever reason), what alternatives such as robust log feeds can offset this blind spot?

   If a PoP already has network or host sensors on it, what will new log feeds tell us that's different from what the sensors can provide?

- **Previous incidents and adversary engagement**

   Even if a small number of assets sit behind a PoP or their mission criticality is fairly low, the SOC may choose to put a sensor there because one or more incidents have occurred on those assets in the past.

   Having focused monitoring capabilities may help analysts run future suspected intrusions to ground much quicker, instead of having to piece together proxy or NAT logs from a firewall or wading through many terabytes of PCAP.

- **Ownership of assets**

   The SOC may have restricted ability to place monitoring capabilities on a network, on the basis of system ownership issues.

   These issues may bring up such situations as (1) comingled government or commercial assets, (2) outsourced or cloud computing, (3) business-to-business (B2B) or government-to-government (G2G) connections, (4) internal or external contracting of IT services, or (5) coexistence of multiple SOCs within a larger organization.

   The SOC may consider partnering with other network operations teams and SOCs to ensure that networks of mixed or ambiguous ownership are monitored by someone.

Across the PoPs where SOC chooses to deploy passive network monitoring, some of these will no doubt include full-session PCAP capture. This choice is also driven by the same factors that drive general passive network monitoring placement. However, it certainly is more constrained by a few factors: long-haul bandwidth to retrieve PCAP and storage for PCAP at the sensor. The choice to record full-session data is often driven by Tier 2's needs to run incidents to ground—where are they getting hit most often, and, therefore, where are network traffic details needed?

9.1.2 Active Network Sensors and Content Detonation

When considering active network sensors, we leverage the same criteria and considerations for passive sensors but with more selectivity, considering the higher costs associated with acquiring them, and their performance limitations. If we had 20 places where a NIDS would make sense, perhaps only two or three really need a NIPS *in active prevention mode.*

The most obvious place where we would want to put a NIPS is where we wish to block direct network-borne attacks, perform bandwidth throttling or "packet shaping," or filter out specific content such as games in social network sites. This would usually be done at an Internet gateway, but major transit or interconnection points within the network or WAN also sometimes make sense. Consider, for instance, being able to block an adversary as it moves laterally across the constituency.

The choice of where to put content detonation devices is perhaps the most straightforward—at any gateways where we exchange email or Web traffic with the Internet or between large networks of differing trust levels. Ideally the constituency has a small number of these, and, therefore, few devices are needed. The SOC may also choose to host its own out-of-band content detonation device in its enclave for the purpose of ad hoc malware analysis.

9.1.3 Purpose–Built Monitoring Devices

Rounding out our conversation of network-based sensors, we address application-specific monitoring and prevention devices such as XML firewalls. Use of these devices is very straightforward. Basically, any system that serves a large number of semitrusted or untrusted users with a corresponding protection technology is a candidate for such a device. For instance, a Web services interface between a government agency and many private corporations' business-to-government (B2G) connections might be a good place for an XML firewall. An externally facing Web server that allows members of the public to access health or financial records might be well served by a database firewall.

While these are certainly intrusion monitoring and defense capabilities, they are also tightly embedded in key applications. Deployment and tuning of these systems should be done in coordination with respective sysadmins.

9.1.4 Host Sensors

Network-based sensors usually get top billing on the SOC's monitoring capabilities list, but they certainly don't provide complete visibility, especially in the presence of obfuscated or encrypted protocols. That said, not every SOC has the money and the time to deploy a full suite of monitoring and prevention capabilities to every host in the enterprise. What should we prioritize? Let's look at what factors should be maximized when considering which hosts to instrument first. (See Table 15.)

Table 15. Host Monitoring Placement Considerations

Maximize	Example(s)
Importance of hosted data and applications' confidentiality, integrity, and availability to mission—possibly expressed in dollars, lives, or impacted users	Key enterprise database servers, billing systems, manufacturing automation control, or supervisory control and data acquisition (SCADA) systems
Number and strength of trust relationships between that system and other hosts, especially hosts residing in other enclaves	Web server directly exposed to Internet Web services systems forming a B2B relationship with a partner company Remote access VPN or webmail servers
Number of, and privileges wielded by, users on that system, especially users residing in other enclaves	Web-enabled financial application server; call center ticketing system
Vulnerability and attack surface exposed by system(s) of concern	Any server that cannot be regularly patched for whatever reason—legacy, operational demands, fragility, and so forth
Stability, maturity, applicability of protection mechanism(s) to that platform	Full HIPS suites for Windows platforms; other components for major UNIX/Linux flavors

Again, it is important not to forget the lessons from Section 8.3.10—not all monitoring tools are applicable to all hosts. In many cases, the most important systems in the constituency are not well suited for a typical HIDS/HIPS suite. We may depend on other tools like configuration checkers, robust logging, and native OS host firewalling to get the job done. It also should be noted that many SOCs that do pursue pervasive host monitoring architectures spend a lot of time identifying, diagnosing, and solving integration issues and high-priority alerts with IT operations and system owners. Prioritizing which alerts to follow up on is key, and the SOC is well advised to take a holistic look at its triaging and escalation process.

9.1.5 Examples

Using the lessons and considerations discussed in this section, let's examine some common network and host sensor placement scenarios.

In our first example, we instrument the main portion of constituency networks, where users perform their regular business computing and access services from the intranet and the Internet. (See Figure 26.)

Let's go over the instrumentation of this network, starting with the Internet gateway at the top left. Here we have a passive sensor or set of sensors that gather SuperFlow data and IDS alerts from each leg of the externally facing firewall. In addition, the SOC may choose to perform sustained PCAP collection on some or all of the traffic on each leg. As an aside, there is a philosophical debate amongst sensor architects as to whether IDSes go on the inside or outside of the firewall. Here, we are doing both, due to the proxying nature of the firewall—putting a sensor on just one side may not tell the whole story. It should be noted,

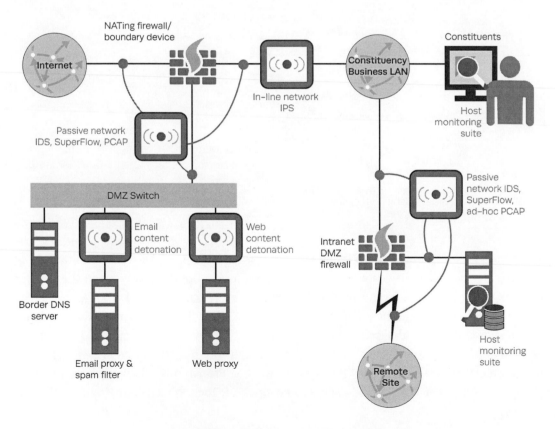

Figure 26. Instrumenting an Internet Gateway

though, that alerts coming from a sensor on the external side of a firewall should be triaged with an understanding that a large portion of scanning and exploit activity probably bounced off the firewall and is of no great concern.

Every constituency border DMZ is shaped differently. In this example, we have email and Web proxies hanging off a leg of the firewall. In the gateway DMZ, we leverage email and Web content detonation to detect zero-day attacks from Web pages and email attachments. The architecture pictured allows us to actually block malicious email and Web content by placing the content detonation devices in-line with their respective proxies.

Moving to the internal LAN at the top right, we have an in-line NIPS looking for any direct attacks inbound or outbound from the general user population. The internal network segment flowing from the firewall is probably the most logical place to put this technology, given its cost and the relative rarity of directed attacks that can be seen on the wire. Ideally speaking, the SOC is also able to instrument desktops belonging to the general host population. Referring to Section 8.3, we will probably leverage a combination of HIDS/HIPS, AV, and antispyware, understanding each of these products' limitations. We may choose to augment this with other packages such as on-demand host forensics and host telemetry data collection.

Chances are, most constituencies also have a significant number of Intranet services that may be found in a consolidated location, possibly behind a firewall. This is an excellent point in the network to look for adversaries traversing the internal network and insider threat. In such cases, it's useful to see the traffic going by this connection with SuperFlow collection and passive IDS. Full PCAP collection may be too costly considering that traffic capture retrieval may not be frequently needed. That said, it's always a good idea to have ad hoc PCAP collection capability on any sensor. If there's anywhere internal to the enterprise that an active or passive host monitoring suite should be placed, it's probably on these servers: intranet Web servers, finance, database, and domain controllers.

Finally, we see a remote constituency site (or sites) hanging off the bottom of the diagram. These are really good places to perform SuperFlow collection and passive content inspection, again, looking for malicious actors moving around inside the network where they're least wary of being seen. In these cases, we also want to have ad hoc PCAP collection tools at our disposal. If there are a large number of satellite campuses, the SOC may consider collecting NetFlow data from border routers in order to maximize their visibility while keeping their IDS appliance count to a minimum (thereby reducing costs).

Moving on to externally facing services, we have the scenario shown in Figure 27.

In this example, the constituency is offering access to sensitive data to external parties such as the general public or another institution such as a business or a government agency. The data is probably being provided through an interactive Web portal and/or

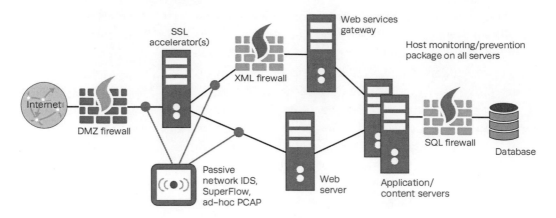

Figure 27. Instrumenting an External–Facing DMZ

machine-to-machine Web services. In either case, there are common elements to the monitoring architecture, so we will address both at the same time.

First, on the left, we have our usual passive IDS and SuperFlow collection at the border of the DMZ. There are several key points to note here. First, we won't bother with NIPS because there are so few protocols flowing in and out of this DMZ that a general-purpose sensor will provide limited benefit. Second, although we are monitoring the network link in between the SSL accelerators and firewall, this traffic is encrypted. Therefore, monitoring is best limited to perhaps inspection of SSL handshakes and connection metadata; PCAP collection here is of almost no use. Third, if we are using a general-purpose IDS, we will probably turn off rules for almost all protocols that are not Web-related. A network sensor geared toward Web traffic monitoring may be a better choice, or the SOC may choose simply to rely on Web server logs and Web application logs and dispense with NIDS altogether. Fourth, if the web server's logs are of sufficient fidelity, SuperFlow collection may not be necessary.

Inside the SSL accelerators, we can leverage an XML firewall to inspect highly structured data flowing in and out of the Web services interface. Similarly, we may also choose to use an SQL firewall in front of a relational database on the backend. Some implementations may feature both capabilities, though this may be seen as overkill. Finally, we will place a host monitoring and prevention package on each server.

9.1.6 Cost

Going back to the Internet gateway example, we have a long list of sensors and log sources we might leverage when looking for or running potential incidents to ground:

- Firewall logs
- Web/email gateway filtering and logs
- In-band NIPS
- Out-of-band NIDS outside firewall
- Out-of-band NIDS inside firewall
- Out-of-band NIDS on DMZ leg
- SuperFlow and PCAP inside firewall
- SuperFlow and PCAP outside firewall
- SuperFlow and PCAP on DMZ leg
- Email and Web malware detonation chamber
- HIPS on every applicable host.

Not all SOCs have the resources to instrument their Internet gateway to such a degree, let alone other parts of the enterprise. Let's look at some techniques for keeping costs related to monitoring coverage under control:

- Prioritize sensor placement using the law of diminishing returns with respect to Table 14 and Table 15.
- FOSS can dramatically reduce acquisition costs, provided the SOC has expertise with these technologies in-house:
  - The most mature SOCs will find a synthesis of commercial, free, and custom tools that meets their needs.
  - Consider leveraging open source tools in deployment scenarios that compose the largest "box count," with expensive tools used only in critical focus areas.
- Operate a modest set of tools well, rather than a large number of tools poorly.
  - Some SOCs are lucky enough to receive significant sta tup funds; this may result in a plethora of different technologies that are expensive to maintain over time.
  - Consider a standard set of "strike packages" that are applicable to most common monitoring scenarios.
- Use each sensor to its maximum potential, without dropping many packets.
  - In large networks, many IDSes sit relatively idle while a handful of sensors are maxed out.
  - Consider consolidating multiple monitoring taps for low-bandwidth connection into one sensor or rely solely on NetFlow and application logs.
  - Properly size sensor platforms to current and projected bandwidth.
- Leverage firewall logs and NetFlow in places where it's not possible to put a network sensor (due to logistics, cost, network ownership, etc.).
- Ensure monitoring technologies match the current and projected threat environment.

- Invest in technologies that match modern and growing threats; as of the writing of this book, these would likely include client-side attacks and activity associated with APT.
- Careful, requirements-driven product evaluations provide insight into internal architecture, allowing the SOC to gauge whether a product and its company will have room for growth, or whether the product is built on a shaky foundation.

9.1.7 Policy

One of the most effective ways for a SOC to ensure comprehensive, mission-relevant monitoring coverage is to have these needs articulated in an IT policy for engineering of new systems and services. The SOC, however, must be very sensitive to how these policies impact constituency resourcing for new and ongoing projects. When authoring policy mandating monitoring coverage for the constituency, consider the following:

- A standard set of "strike packages" helps support judicious use of policies.
  - Cut-and-dried directives like "deploy this package on XYZ systems" can work in some cases (e.g., Windows desktops and servers) but much less so in others.
  - Mandating the deployment of a specific tool, especially on the host, can become onerous with larger constituencies, which is the case with coordinating SOCs.
  - A good alternative is to mandate that certain capabilities exist and provide enterprise licensing for a tool that meets the need. In this way constituent programs and subordinate SOCS can use their own, if they so choose.
- One of the best ways to incorporate network-based monitoring is to require new and expanded projects to coordinate with the SOC to assess needs and implement the right level of monitoring.
  - This helps the SOC work with system owners to understand their mission and tailor capabilities accordingly.
  - This also helps the SOC build credibility with constituents and ensures that all parties are aware of appropriate escalation contacts and processes.
- It is simplest for SOCs to maintain their own budget supporting comprehensive monitoring coverage and staffing.
  - It helps the SOC advertise its capabilities as a service and documents its cost savings to other programs and projects.
  - Moving to a fee-for-service model or "tax" can become challenging: many programs and projects will not budget for new capabilities, and the SOC's year-to-year budget planning will become overly complex or subject to third parties that the SOC will have a hard time influencing.

- Some SOCs will look to especially large projects to help fund monitoring tools at initial deployment time and will build in recap and staffing costs for outyears.
- Larger coordinating or tiered/distributed SOCs may formulate a standard memorandum of agreement (MOA) or memorandum of understanding (MOU) to formally recognize tailored monitoring capabilities for certain enclaves.
  - This helps document monitoring duties, technical POCs, and management POCs in the long term.
  - Constituency programs will often feel better served when their specific monitoring needs are codified in some sort of documented format, versus informal agreements.
  - The SOC can establish expectations for what system owners will receive in return, such as regular cyber SA reporting; more than a "we'll call you when we see something bad" is helpful.

9.1.8 Virtualization and the Cloud

Our discussion of how to instrument constituent systems assumes that they are hosted internally to the constituency. Moreover, traditional instrumentation assumes bare-metal installation of hosts and services, where one entity owns the entire computing "stack." As of the writing of this book, there is a massive shift to virtualized infrastructure, and many services are being pushed into the "cloud" [233].

In the case of virtualization, not that much changes in terms of how to instrument the network. There are a few issues to look out for, however. When we introduce hardware virtualization (e.g., VMware ESX/ESXi) into the environment, we also end up virtualizing a significant portion of the LAN architecture, whereby VMs may talk to one another without ever traversing a physical switch or router. This can present a challenge because most active and passive IDS/IPS technology depends on having a physical network segment to copy network traffic from, or to insert a device into. Luckily, several commercial vendors now have virtual IDS/IPS appliances that can be inserted directly into the virtualized infrastructure. Or, if a SOC wants to go the FOSS route, Snort and Argus could be loaded onto a VM that lives in the server farm.

Three challenges remain. First, performing gigabit (or greater) PCAP and traffic metadata collection entails nontrivial storage and performance that the virtualized environment may not readily support. Second, the virtualized sensor platform may require specific configurations and control not normally granted to tenant organizations such as the SOC. Third, the confidentiality, integrity, and availability of those sensors may be harder to guarantee when resident in a third-party hosted environment. When implementing sensor platforms in a third-party environment, the SOC will likely need to work with its infrastructure provider to ensure its unique needs are met.

Moving on to cloud computing, instrumenting our networks becomes much more challenging. In all common scenarios (platform-, software-, or infrastructure-as-a-service [PaaS, SaaS, IaaS]) the landlord, not the tenant, controls the network infrastructure. As a result, the SOC's options for use or placement of sensor technologies may be severely limited. Sometimes the landlord/cloud provider will include security monitoring feeds as part of its service or as an optional capability. This can be problematic if the SOC has its own platform of choice, which is usually the case. While the cloud provider may agree to provide the SOC alerts from its own IDSes, the SOC may not be able to influence tool choice or signature tuning, which usually is a requisite for most SOCs to make good use of an IDSes. In some cases, the best option for the SOC may be to enhance its log collection from systems moved to the cloud. In the case of IaaS, the SOC should include some sort of HIDS/HIPS packages as well.

Alternatively, the SOC's parent organization may wish to outsource CND for the portion of its enterprise that exists in the cloud to the cloud owner or its designated CND provider. This arrangement often works best for IT assets that can be somewhat cleanly separated from the rest of the constituency (e.g., an Internet-facing Web presence that serves largely static content). With such an arrangement, there is comparatively little ambiguity between the roles of each SOC. Nevertheless, it's important for the constituency SOC to establish clear lines of communication and service agreements with the cloud-provider SOC.

As an aside, effective instrumentation is only one challenge for the SOC when dealing with outsourced or cloud assets. When the SOC's parent entity does not own the entire computing stack, aspects of control and response become much more challenging. If a host is infected, can it be taken offline quickly? Can it be imaged? Even if the answer is yes, what SLAs are in place to support this? In order to support better defense, the SOC should pursue opportunities to influence how IT is outsourced to the cloud whenever possible.

9.2 Selecting and Instrumenting Data Sources

One of the most frequent questions posed by SOCs is, "What log data should we gather?" Every constituency is different, so, while there are many common themes we will discuss, the answer always varies. This section captures these commonalities and presents a pragmatic, operations-driven approach to prioritizing what data to gather.

9.2.1 Choosing the Right Data Sources

In the modern enterprise, there are several drivers for collection of IT security log data:

- Computer network defense
- Insider threat monitoring and audit collection

- Performance monitoring
- Maintenance troubleshooting and root-cause analysis
- Configuration management.

While all of these missions have overlap both technically and organizationally, the SOC is often at the nexus of log collection since it needs to pull from so many different sources. It is easy to get buried in log data, so it is important for the SOC to take a concrete, requirements-based, value-driven approach to collection, use, and retention.

Let's look at some key questions, understanding they will drive not only what data we collect but which systems we collect it from, whether we filter or deduplicate it, and how long we retain it—in what format and under what circumstances:

- What is the mission of the systems being considered for monitoring?
  - Each monitored system's link to the mission
  - Mission criticality (monetary, lives, etc.).
- What trust is placed in users of the system(s) and hosted services?
- What is the assessed or perceived level of integrity, confidentiality, or availability of the system(s), data, and services?
- What is the perceived and assessed threat environment? How exposed are systems to likely adversaries?
- Are the systems (or their audit data) under any sort of legal, regulatory, audit-related, or statutory scrutiny, outside of those placed on it by the SOC?[1]
- What quality, visibility, and attack life cycle coverage is offered by host or network CND sensor instrumentation (IDS/IDS) on or near hosts in question? Is that enough, or do we need logs as well?
- Quality and clarity of logs produced:
  - Human and machine readability; does the log use cryptic messages and obscure encoding, or can a human easily read it in a consistent American Standard Code for Information Interchange (ASCII) format?
  - Does one log entry correspond to one logical event, or is human or machine correlation needed to piece together several disparate audit records?
  - Are the clocks of all systems considered for monitoring synchronized?

---

1 This can be politically sensitive. The SOC may be compelled to collect logs for regulatory reasons, even though the SOC doesn't otherwise need them. The SOC is encouraged to take an objective look at the impact of policy and procedure on log collection, ensuring it does not place undue burden on SOC resources. Audit log policy is a topic often dominated by guesses and innuendo and not so much by facts and good judgment. Just because someone has to collect and retain a log doesn't mean it has to be the SOC.

- Do the events include details of what actually happened (e.g., extra data fields and human attribution)?
- Are system owners willing to adjust their logging polices to meet SOC needs?
- Availability of logs:
  - Are logs written in an open format that is vendor neutral, and/or can they be interpreted out of the box by audit collection tools or SIEM currently in use?
  - If no existing parsers exist, what resources are needed to make appropriate use of them within their intended data aggregation or analytic framework?
  - Are system owners willing to provide direct or mediated access to log data in their original format and in real time?
  - What is the overall volume of logs being written? Will the networks and systems that connect the source system with the master SIEM or log aggregation node support the requisite bandwidth and disk space? Could the logs be scheduled for batch transmission and input at off hours, and, if so, can the correlation engine cope with this delayed ingest of data?
- Coverage that the logs provide:
  - Will a given log feed cover a wide portion of the constituency, such as with Windows domain controller or Web proxy logs?
  - Or will the logs provide enhanced coverage for a specific high-value application (e.g., a financial management system) that deserves deep visibility and detailed monitoring use cases?
- Collection and analytic framework:
  - Does the SOC have an existing log aggregation or SIEM tool in place?
  - If the tool can accept said log type, does it provide the analytical framework and correlation tools necessary to make good use of the data?
  - Will introduction of this new log data into the existing collection and analytic framework dilute the quality of the logs already being gathered? Will it severely slow analysts' ability to access the most important log feeds?

Make no mistake about it, when it comes to enterprise-grade audit collection, the devil is certainly in the details. Many enterprise audit efforts become infamous for their high resource cost and low perceived return on investment. It bears repeating: the SOC must carefully evaluate, tune, and retune every log feed it collects in order for its efforts to be worthwhile.

Here's one way to get started. Pick a handful of use cases or threats the SOC wishes to detect, pick which feeds are necessary to build out those use cases, and then put them into operation. There are several virtues to this approach: (1) the SOC has a clear set of requirements and goals to meet, (2) there is a concrete set of deliverables and returns that will result, and (3) the set of data being collected is seen as finite and therefore manageable.

Another challenge for SOCs when working with security-relevant logs in support of incident response is user attribution. Due to the very nature of the ways many OSes and applications function, it is difficult, or sometimes nearly impossible, to tie a user to every action that a system carries out. For instance, many actions carried out by a host may occur at the system level and are not user initiated. In other cases, the identity of the user is not carried throughout the length of the system. Such is the case, for instance, when attributing who inserted a USB stick into a desktop computer or understanding what SQL database query is linked to which application user.

> **User attribution is hard. Attaining probable or confirmed connections to people's actions is a big part of running incidents to ground.**

We think we saw an attack originate from outside the enterprise. While its originating IP netblock is assigned to Kazblockistan, is this the location of the attackers or just the next hop out? You may have logs that say Alice printed a pile of sensitive documents that ended up at her house weeks later, but was she the one who actually walked out the door with them?

True user attribution means you know beyond the shadow of a doubt that someone's actions are actually tied to what shows up in an IDS alert or system log. Inside the enterprise, this can be a challenge, even with clock-matched security camera video and copious system logs. Outside the enterprise, this is nearly impossible. This is why, when it comes to actually prosecuting users for their misdeeds, SOCs work with law enforcement and legal counsel.

9.2.2 Typical Sources

We have established the need for concrete requirements and the need for careful attention to detail when considering which log types to collect. In this section, we cover the most common data feeds that a SOC will leverage. In Table 16, we provide an overview of the many common log sources collected by SOCs. This table includes information about (1) what their data reveals at a cursory level, (2) a rough order of magnitude of event volume, (3) what their volume depends on most heavily, (4) what the general value of their data is, and (5) common fields of interest. Before we elaborate on the table itself, there are a number of assumptions and caveats that should be recognized:

- Some of the most popular data feeds for Small and Large Centralized SOCs from Section 4.1.2 are bolded; again, understanding there is no "one size fits all."
- The number of devices listed assumes a typical constituency of 5,000–50,000 users and IPs, with a centralized SOC organizational model in mind.
  - The number of actual hosts in a given constituency will obviously vary. The numbers listed in this table are given to help the reader form an estimate of data volume—given the number of devices and the typical number of events per device.
  - The number of devices refers to the number of end systems generating this data, not systems that serve as the collection point. For example, we list tens of thousands of devices for AV, even though their data is likely rolled up to one or a few AV management servers.
- In addition to the items listed in "Volume depends on," the volume of every feed depends on the following four items:
  - The volume of activity seen by the device generating the events
  - How that data feed is tuned at the generating device and upstream
  - The detail and verbosity available from the given source devices
  - How the end device(s) are configured.
- "Subjective value" gives a sense of the likely quality of each particular data type, assuming it has been tuned properly.
  - The actual value of a given data feed will, of course, vary within the context of the constituency, implementation, mission, and particular product implementation.
  - Its value is described as follows:
    - 4 = Excellent
    - 3 = Good
    - 2 = Fair
    - 1 = Poor
    - 0 = None/Not applicable
  - We show the value of a given data feed under the following three circumstances:
    - "Tip-off" means the data will help direct the analyst's attention, without additional enrichment from correlation by a SIEM.
    - "Tip-off with correlation" means a given log entry will provide a good incident tip-off, assuming it is enriched or correlated with other data.
    - "Supporting context" means it helps an analyst establish the ground truth of an incident.
- "Fields of interest" describes the specialized fields commonly found in that data source, which are of particular interest for correlation or forensics purposes. We assume the following standard fields in each data type:

- Date and time, at least down to the second, possibly with time zone
- Source IP, port, and hostname, if applicable
- Destination IP, port, and hostname, if applicable
- IP and/or hostname of the device that originated the event
- Event name, possibly with a detailed description.

For a more detailed illustration of various log types, see Chapter 2 of [48], and [234]. For an alternative perspective of which logs to collect, see page 22 of [235].

9.2.3 Tuning

We have selected a data feed for collection—how should we pare down its volume? Recall our discussion of data quality and tuning, from Section 2.7, and the tools we have at our disposal to gather, correlate, and display data, from Section 8.4. There are two classic approaches that SOCs may take in selecting and tuning data sources: tune up from zero or tune down from everything. Table 17 identifies the pros and cons of each approach.

Luckily, modern SIEM and LM systems give the SOC a great deal of leeway in how to decide which data to collect and retain, allowing us to pursue a hybrid of these two approaches. For instance, the SOC can collect all of the data being generated by firewalls and Web proxies but only use a very small percentage of those feeds for purposes of correlation with, and display to, analysts. Great care must be taken not to overburden collection systems or analytic frameworks with too much data. Refer to Figure 8 in Section 2.7. There is an optimum spot for every data aggregation framework, where we strike a balance between performance and volume.

There is a common pitfall when defining audit policies—generating messages only on a "fail" but not on a "success." Failure events include users typing in the wrong password or being blocked from visiting a website. Failures mean a device did its job. It stopped someone or something from doing what it shouldn't do—this is a good thing. Successes—file modification granted, file transfer completed, and database table insert—are often where we actually need to be concerned. This leads us to an important point:

> **Don't log just the "denies"; the "allows" are often just as important.**

Consider situations in which "allows" are often more important than "denies"—malware beaconing, RATs, data exfiltration, and insider threat. With only failure attempts logged, we may not get the whole story. Also, when we take notice of the "allows," we have events that we need to take action on; "failures" usually require no further action. Failures may tip off an analyst to activity of concern or trigger a correlation rule; successes are just

Table 16. Comparison of Common Security–Relevant Data Sources

What	What it tells you and use cases	Typical # of events per day per device	Volume depends on	Typical # of devices	Subjective Value As...			Fields of interest
					Tip-off (raw)	Tip-off w/ correlation	Supporting context	
NIDS/NIPS[1]	Attacks seen traversing the network; see Section 8.2.2	100s–100,000s	Signature set, features	10s–100s	3	4	1	Allow/block, vulnerability ID
NetFlow[2]	High-level statistics on all network traffic seen; see Section 8.2.3	100s–1,000,000s	Location of tap	10s–1,000s	1	4	4	Bytes in/out
Full network session capture (PCAP)	Full details of entire network conversation; see Section 8.2.5	N/A[3]	PCAP collection filters	1s–100s	0	0	4	Everything about each layer of the protocol stack
HIDS/HIPS	Attacks, other activity seen at the host; jack-of-all-trades, master of none; see Section 8.3.3	10s–1,000s	Tuning of signature set	1,000s–10,000s	3	4	2	Process name/ID, action taken (allow/block), file name

1 Recall our discussion from Section 8.2—an analyst needs full PCAP, signature definitions, and NetFlow to make heads or tails of an IDS alert.

2 Recall our conversation in Section 8.2.4—data from these tools can often support many of the same uses cases as firewalls, and vice-versa. Some NetFlow collection systems also produce metadata on popular protocols such as HTTP and SMTP, making Web content filter and email gateway logs somewhat redundant.

3 PCAP doesn't get delivered in events; however, it generally dwarfs all other data sources listed in the table, in terms of volume. For more details, see Section 8.2.5.

Table 16. Comparison of Common Security–Relevant Data Sources

What	What it tells you and use cases	Typical # of events per day per device	Volume depends on	Typical # of devices	Subjective Value As...			Fields of interest
					Tip-off (raw)	Tip-off w/ correlation	Supporting context	
System integrity checker	Detailed changes to key host configuration files, settings; see Section 8.3.5	10s–10,000s	Volume of automated admin tasks	1,000s	3	4	2	File name, hash
AV	Hits on known, easily identifiable malware; see Section 8.3.2	10s	Exposure to downloaded files	10,000s	4	4	1	File name, virus name, user name, action taken
User activity monitoring	Everything a user does while logged in; see Section 8.3.8	100s–10,000s	# of users under scrutiny	10,000s	1	4	4	User name, action taken
Intellectual property and DLP	Attempts to infiltrate or exfiltrate documents and other information from the enterprise	1s–100s	Policies on removable storage, file transfer use	10,000s	2	4	4	User name,[1] removable storage unique ID, protocol or device used
Network firewall	Activity and bandwidth seen across firewall, NAT records, and possibly FTP and HTTP traffic details	100,000s	Firewall rule set, volume/ diversity of traffic	10s	1	4	3	Session ID for NATing, bytes in/out

1 Many DLP events will not be labeled with the user name of the person plugging in or removing USB mass storage devices. This is a good example of how user session tracking in the SIEM might provide a compelling use case, albeit a potentially complicated one.

Table 16. Comparison of Common Security-Relevant Data Sources

What	What it tells you and use cases	Typical # of events per day per device	Volume depends on	Typical # of devices	Subjective Value As...			Fields of interest
					Tip-off (raw)	Tip-off w/ correlation	Supporting context	
Web content filter	Details of all proxied traffic, usually HTTP; tracking Web usage, malware sites	10,000s–1,000,000s	Filtering rules in use	1s–10s	3	4	3	Allow/block, URLs, referrer, user agent, Web site category (news, shopping), website reputation score
Mail gateway	Details of email that goes in and out of enterprise; insider threat, data leakage	1,000s–100,000s	Quantity of spam	10s	1	4	3	To/from address, subject, attachment name, allow/block/quarantine
Email server logs	Details of ALL email that server handles	1,000s–100,000s	Amount of internal email traffic	1s–10s	1	4	3	To/from address, subject, attachment name
Content detonation device	Malware found in streams of Web or email data	10s–100s	Amount of malware ingress to the network	1s	4	4	4	Malware name, file name, source content, hash, vulnerability ID
Router/switch	Link, port up/down, router changes, location of MAC addressed attached to network	100s–10,000s	Verbosity of logging level enabled	100s–1,000s	1	2	2	Bytes in/out, MAC

Table 16. Comparison of Common Security–Relevant Data Sources

What	What it tells you and use cases	Typical # of events per day per device	Volume depends on	Typical # of devices	Subjective Value As...			Fields of interest
					Tip-off (raw)	Tip-off w/ correlation	Supporting context	
DNS	Major events on the DNS server such as zone transfers; DNS requests from internal servers can reveal malware beaconing.	10,000s–1,000,000s	Verbosity, DNS caching in place	1s–10s	0	4	2	Contents of DNS query and response
Dynamic Host Configuration Protocol (DHCP)	Records of DHCP lease requests/renewals; what systems were on the network, when, and where	100s–10,000s	DHCP lease timeout	10s–1,000s	0	1	2	DHCP lease info, lease MAC
NAC	Results of any system attempting to gain logical access to the network	100s–10,000s	Complexity of policy, openness of network	10s–100s	2	4	3	System details (OS, patch level), MAC, allow/quarantine
Remote Access System and VPN	Attempts to gain remote access to the enterprise	10s–1,000s	Size of remote worker pool, # of partner orgs	1s–10s	0	3	3	User name, remote IP, client version

Table 16. Comparison of Common Security–Relevant Data Sources

What	What it tells you and use cases	Typical # of events per day per device	Volume depends on	Typical # of devices	Subjective Value As...			Fields of interest
					Tip–off (raw)	Tip–off w/ correlation	Supporting context	
Windows Doman Controller[1]	Authentication, access control events for all systems on domain;[2] variety of use cases with careful interpretation	10,000s–100,000s	Audit policy	1s–10s	2	4[3]	3	User name, user privileges, success/fail, other Windows specifics
Local Windows event logs	Authentication, access control events for that particular host, regardless of whether it uses domain privileges	10s–100s	Audit policy	10,000s	2	4	4	User name, user privileges, success/fail, Other Windows specifics
Single sign-on (SSO) and identity access management	Consolidated tracking of logical user access to enterprise resources, common user names spanning disparate systems	1,000s–100,000s	Number of systems SSO-enabled	1s–10s	0	3	4	User name, translated (real) identity, user attributes

1 Logging activity on the domain can be fairly challenging due to the complex nature of how Windows records logs, at least in part because of the complexities inherent in the Kerberos protocol [318].

2 It's important to recognize that activity on systems in the domain that occurs under a nondomain account often won't get rolled up to the domain controller. This is very important if an attacker uses a local system account to do something bad and you were expecting to see that information show up in the domain controller logs. However, domain controllers usually serve a logical consolidation point for logs, whereas instrumenting each end host, even just servers, can be costly. If used properly, domain controller logs can be a smoking gun for spotting insider activity.

3 Unfortunately, the way Windows writes user login/logoff records makes tracking user sessions challenging. See [319] and [320].

Table 16. Comparison of Common Security-Relevant Data Sources

What	What it tells you and use cases	Typical # of events per day per device	Volume depends on	Typical # of devices	Subjective Value As...			Fields of interest
					Tip-off (raw)	Tip-off w/ correlation	Supporting context	
Physical access control (badge reader)	Physical access to enterprise facilities (badge in, possibly badge out); insider threat	1s–100s	Penetration of deployment, requirement for each person to swipe in and out	1s–100s	0	3	3	User ID, room #
UNIX/Linux OS logs	Privileged system actions (usually automated activity)	10s–1,000s	Amount of system activity	10s–1,000s	2	4	4	Process name/ID
COTS application, custom-built apps	Application-specific actions, logical user access and changes to objects and data; goldmine for insider threat monitoring; account compromise and data leakage	10s–100,000s	User population, application type and complexity	1s–100s	?1	4	3	User name, action, object name

1 Custom and COTS applications, which are often Web-enabled, offer very interesting opportunities for monitoring because they often closely support core missions of the constituency. However, they often write logs in a variety of formats, requiring custom log parsers and human interpretation. A SOC that has resources to leverage even a few of these can really hit a home run when they uncover malicious activity that manual human review couldn't.

Table 16. Comparison of Common Security–Relevant Data Sources

What	What it tells you and use cases	Typical # of events per day per device	Volume depends on	Typical # of devices	Subjective Value As...			Fields of interest
					Tip–off (raw)	Tip–off w/ correlation	Supporting context	
Web server	Results of HTTP requests to all pages on server	100s–100,000s	Complexity of website, number of users, popularity	1s–100s	1	3	3	URL, HTTP code, referrer URL, user agent, user name (if authenticated)
Database	Some or all SQL commands and possibly part of their results	1,000s–1,000,000s	Complexity of database, number of users and applications	10s–1,000s	1	3	3	SQL statement issued, database user and privileges
Vulnerability scanner	Known vulnerabilities, services, and other details about scanned hosts; see Section 8.1.3	10,000s	How often you scan	10s	1	2	3	Vulnerability name, scan date, other

as important in understanding the full story. Moreover, "allows" are ripe for the kind of analytics and correlation SIEM tools bring to bear.

9.2.4 Obtaining Feeds

Actually getting data feeds hooked up and turned on can be a big obstacle for the SOC, often due to political hurdles rather than technical ones. There are several strategies the SOC can pursue to overcome these challenges.

First, the SOC is advised to pursue a consistent, aboveboard process in acquiring new data feeds. While quick and informal agreements with system owners and sysadmins can get results quickly, they may not be durable, due to personnel turnover. If the constituency has an engineering or CM process that supports timely delivery of services, the SOC should leverage this existing process for articulating requirements.

Second, in medium to large constituencies with Centralized and Tiered SOC models, a formal MOA or MOU may help when setting up a major set of data feeds or targeted monitoring engagement. This memorandum articulates several items:

- What data is being gathered
- What it is being used for, such as targeted use cases, general CND, insider threat, and so forth

Table 17. Approaches to Tuning Data Sources

Approach	Pros	Cons
Start with the entirety of a given data feed and tune down to a manageable data volume that "meets common needs"	Requires little foreknowledge of the data being gathered Easiest to implement Enables SIEM/LM tools to leverage full scope of data features and event types offered	May overwhelm tools and analysts if data feed is too voluminous If methodology is used for many data feeds, poses exponential risk of "data overload" "Default open" filtering policy toward data collection poses long-term risk to data aggregation systems as feeds change over time
Start with a candidate data feed, and tune up from zero, focusing only on what is deemed useful or important	Keeps data volume low Focuses systems and analysts only on what is deemed to be of interest Less problematic for shops just getting started	Carried to its extreme, limits value given time/effort granting SOC access to given data feed Analysts blind to features of data feeds not explicitly set for input into data collection systems Approach may require more labor to implement

- Who is responsible for formal audit review and long-term log retention
- Whom to call if the data feed goes down or changes in any major way, or to decide whether escalation is needed for an incident detected from one of the supplied data feeds
- How the collected data will be secured, especially if any steps need to be taken to protect users' privacy or data confidentiality
- Reference to any important authorities such as a SOC charter or CONOPS
- Any additional expectations of the SOC, system owners, or sysadmins.

Third, the SOC will benefit immensely if the constituency establishes the SOC as the CND provider of choice for the constituency and, potentially, the preferred audit data collection hub. This makes the SOC the default recipient of audit data for all new systems and should compel new programs and projects to proactively approach the SOC. At the very least, the SOC should have policy formally recognizing it as having the right to ask for and retain audit data from constituents.

Fourth, it is important to recognize the technical impact audit data collection places on constituent systems. There are several tips we can offer here as well:

- Minimize the number of agents deployed, especially on end systems. If the SOC can completely avoid placing an agent on a constituent system, the SOC will avoid blame for any sort of impact when a system goes down or is performing slowly.
- Carefully tune performance-related parameters of the agent, such as the polling frequency for events, and the number of alerts retrieved in each poll.
- Leverage existing collection points (such as syslog aggregation points and management servers) where they exist, provided they meet the following criteria:
  - Data is delivered to that collection point without substantial loss in original fidelity and detail.
  - The SIEM or LM system has an agent for the collection point.
  - Data is delivered in a timely manner so it does not disrupt the correlation capabilities of the SIEM; this usually means delivery of events in less than five minutes.
- Leverage assured delivery where such options exist:
  - Connectionless protocols (such as syslog over UDP) can be a problem because events can be lost in network congestion or lossy links—consider instead syslog over TCP.
  - Placing the agent close (logically or physically) to the source systems may help since the SIEM or LM system may offer encrypted, assured delivery.
- Consider using a SIEM or LM system that can transmit events from one agent to multiple destinations, thereby supporting redundancy and COOP, if needed.

Finally, we should recognize that the SOC will likely want to collect data from systems or applications for which its SIEM or LM system does not have an agent. The constituency will almost certainly have applications or systems that record security-relevant audit logs that do not follow a recognized format. As we discussed in Section 8.4.5, several audit data "standards" exist, but none have gained universal adoption.

9.2.5 Long-Term Maintenance

IDS sensors and audit data feeds require long-term care and tuning. For SOCs with large constituencies and a variety of data feeds, this can be a daily battle. Systems are constantly being installed, upgraded, migrated, rebooted, reconfigured, and decommissioned. All of these events have the potential to present blind spots in the SOC's monitoring coverage. Constant vigilance is needed by SOC sysadmins to ensure this does not become a serious problem. Here are a few tips to keep things from getting out of hand:

First, enforce robust but not overbearing CM for SOC monitoring data feeds. With this, sysadmins can keep track of changes to monitoring systems. Some SOCs choose to maintain a list of systems from which they should be getting logs. This list can be scrubbed from real-time data feeds through manual or automated means such as scheduled reports or correlation rules in the SIEM. SOCs are also advised to maintain a technical POC list for all data feeds, as they do for NIDS/NIPS sensors placed at remote sites.

Second, maintain regular contact with constituency sysadmins and engineering process change boards, to keep track of changes to systems. Having regular representation at the boards to maintain awareness of new projects may help.

Third, check data feed status, either daily or every shift. Just because an agent is green doesn't necessarily mean the data feed is online. It may just mean the agent software hasn't crashed. Consider performing regular checks against feeds from high-value targets to ensure there are no interruptions. Also, perform regular checks against SIEM content that is dependent upon key data feeds—is a dashboard blank because there are no attacks today, or because the feed is down? Either is a strong possibility. To keep a more real-time view on health and welfare, consider making this an hourly duty for junior sysadmins or Tier 1 analysts.

One virtue of maintaining vigilant watch over data feeds is that feedback to end system owners will help them recognize that, yes, the SOC is indeed watching. It is doubly important to stay vigilant, because not only does the SOC want to minimize downtime in event feeds, but if outages are caught in real time, system owners can be contacted and asked, "Hey, what did you just do?" This will minimize the time needed to track down the changes that caused the outage.

9.2.6 Data Retention and Leveraging Data Feeds for Audit

The SOC's log collection architecture can often be used to support log audit review, especially in cases where such review is mandated by law or regulation. There are many advantages to fusing audit and CND data collection efforts—the SOC will likely have access to a large set of audit data as it is the *de facto* collector, and the logs can be brought into one place while serving more use cases. On the other hand, this may become a burden to the SOC if not executed carefully. Here are some tips to consider:

First, the SOC's mission does not include full-scope audit review; security officers (in government, ISSOs) and sysadmins do that. SOCs almost never have the resources to perform comprehensive, widely scoped reviews of the constituency's torrent of audit logs. The SOC is there to perform targeted or enhanced correlation and monitoring. This expectation and division of responsibilities must be made clear to security and compliance stakeholders as well as system owners and admins.

Second, those who are granted access to audit logs (e.g., ISSOs and personnel within the office of the CIO) should be granted access to the slice of logs and roll-up reports necessary to fulfill their job. Unfettered access to all logs by a widespread group of people outside the SOC will inevitably lead to conflicts in incident identification and escalation. It will also risk compromise of sensitive insider threat cases. The SOC should carefully work to avoid letting others move too far into the CND area of responsibility.

Third, collection and retention of the constituency's full-scope audit log data will require additional resources—both for systems and the people to manage them. The SOC is strongly advised to procure additional resources in order to sustain this activity.

Fourth, the SOC may be able to leverage existing tools and infrastructure to get an audit pilot off the ground. However, the SOC should consider an audit collection architecture that utilizes a scalable, tiered approach. By segregating data not used for CND purposes, the SOC will minimize the impact on its own operations. For instance, SIEM or log aggregation agents/collectors can be used to extract audit data once and transmit it to separate SIEM/LM instances.

Finally, by taking on a portion of the audit mission, the SOC's log collection systems will be subject to collection and retention requirements that may exceed those driven by the CND mission alone. In the world of audit requirements, technical challenges related to data storage volume and cost are often ignored. The SOC must proactively manage these requirements, because they will impact performance of SOC tools and personnel.

Table 18 suggests guidelines for online log retention within the SOC, recognizing the needs of SOC Tier 1 and Tier 2 analysts, as well as the need for external audit and investigation support. These can be used as a *starting point* for the SOC to evaluate how long it believes it needs to keep data readily available for query, in the context of its own operations and mission.

Table 18. Suggested Data Retention Time Frames

What	Tier 1	Tier 2+	External Support
IDS alerts and SIEM–correlated alerts	2 weeks	6 months	2+ years
NetFlow/SuperFlow logs	1 month	6 months	2+ years
Full-session PCAP	48 hours	30–90 days	2+ years
Audit logs	48 hours	6 months	2+ years

The most common standard audit retention policies, at least those in government agencies, usually set audit data retention at 60 months or more. These stand in contrast to how long SOCs usually must retain data for their own purposes. The most onerous requirement usually stems from supporting external investigations, where a law enforcement entity may approach the SOC and ask for logs on a given subject as far back as it has them, possibly for several years. This can be a real challenge. Consider full-session PCAP. This is not audit data, but keeping it around for case support is definitely beneficial. On the other hand, retaining PCAP for several months or more can be extremely expensive with high throughput connections. Extracting data on a given subject from several years of log data can be very laborious. It is something few SIEM and LM products excel at, in part because user attribution is so challenging.

Chapter 10

Strategy 8: Protect the SOC Mission

■ .... ■

Operating a SOC presumes that at some point the constituency will be compromised. Moreover, actors of concern include individuals with legitimate access to constituency IT resources. Following this logic, the SOC must be able to function without complete trust in the integrity, confidentiality, and availability of constituency assets and networks. While the SOC must have strong integration with constituency IT systems, it must be insulated from compromise. In our eighth strategy, we examine ways to keep the SOC's information and resources out of the hands of the adversary, while maintaining operational transparency.

Military and civilian intelligence organizations must closely protect their sources and methods in order to sustain their mission. Were adversaries to discover how and where they were being watched, they would instantly gain the ability to circumvent detection. The relationship between the SOC and a cyber adversary is no different. If attackers knew where sensors are placed and what signatures are running on them, they would be able to craft an attack and persistence strategy that would go unnoticed by the SOC.

> **A SOC is able to execute its mission precisely because the adversary does not know where or how monitoring and response capabilities operate.**

Even the best SOCs have gaps in their visibility. Moreover, knowledge of what monitoring tools are in use allows the adversary to mount direct attacks against them or, more often, shape its attacks to avoid detection. Best of breed SOCs operate some or all of their systems in an **out-of-band** fashion that isolates them from the rest of the constituency. In order to protect the SOC mission, our design goals are as follows:

- Achieve near zero packet loss at designated monitoring points of presence.
- Prevent the adversary from detecting the presence of (and evading) monitoring capabilities such as IDS and IPS.
- Ensure delivery of 100 percent of security events from end devices to SOC monitoring systems while protecting them from prying eyes, when necessary.
- Support the survivability of the SOC mission, even when portions of the constituency are compromised, and prevent unauthorized access to SOC assets.
- Protect from disclosure sensitive documents and records maintained by the SOC.

We discuss how to address these goals in a bottom-up approach, starting with isolating IDSes and ending with considerations for data sharing. One thing to keep in mind—being overly cautious can get very expensive and doesn't necessarily help the constituency's perception that the SOC is the proverbial Big Brother. There's a fine line between good IT hygiene and paranoia, and that line is different for each SOC.

10.1 Isolating Network Sensors

In this section, we briefly cover common methods for redirecting copies of Ethernet traffic from constituency networks to the SOC's sensors. As these topics are covered extensively in existing literature, we summarize some important points related to reliability, cost, and protection of the sensor in question.

Figure 28 illustrates popular approaches to making copies of network traffic.

Starting at the top left, we see a **network hub** that is being used to copy network traffic to a passive network sensor. This is the most straightforward approach. By inserting a layer 1 Ethernet device between Alice's and Bob's networks, the sensor will see a copy of the traffic passed. However, there are a number of reasons this is less than desirable. Packet collisions and packet latency can become a serious problem. By using a hub, packets will be dropped and the sensor will miss traffic. In addition, most modern networks operate at 1 Gb/s or 10 Gb/s speed; hubs essentially do not exist in speeds faster than 100 Megabits/s. Finally, hubs are generally not very fault tolerant. Thus, network owners are unlikely to approve the placement of a flimsy device between two networks.

At the top right, we replace the hub with a layer 2 or layer 3 network switch. This switch is configured with a switched port analyzer (SPAN) to copy or **"span"** traffic from one or more source ports or virtual LANs (VLANs) to the port hosting a network sensor.

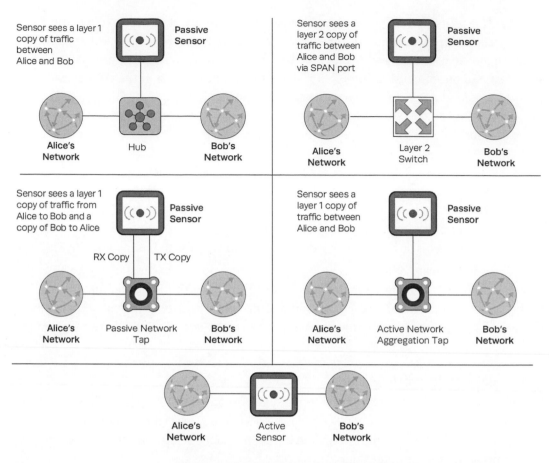

Figure 28. Copying Network Traffic

This approach offers most of the benefits of the hub approach, by using enterprise-grade network equipment that probably is already in operation across the constituency. The major caveat to this approach is that the SPAN must be set properly, and that its configuration must continue to match the intended source ports and VLANs down the road. As network topology changes, respective SPAN configurations must be altered to match.

In the middle of the diagram, we see two different scenarios that leverage a network tap. A **network tap** is essentially a device inserted between two network nodes that makes a copy of all network traffic flowing between them. On the left, we see a passive network tap (in the case of copper) that simply makes an electrical copy of the traffic flowing, or, in the case of optical taps, uses a mirror to actually split the transmit (TX) and receive (RX) light beams. With "dumb" or passive network taps, we must use a sensor that has built-in logic that recombines the RX and TX lines into one logical stream of traffic suitable for decoding. On the right, an active network aggregation tap does this work for us but, at the same time, has the disadvantage of saturating its monitor port if both Alice's and Bob's aggregate bandwidth exceeds that of the monitor port. Popular manufacturers of network taps include NetOptics [236], Fluke [237], Network Critical [238], and Datacom Systems [239]. Network taps aren't generally subject to the same range of misconfigurations that switch SPANs are; that said, network sensors certainly get disconnected from time to time.

> **Keeping SPAN configurations up–to–date so network sensors don't go "dark" is a major pain point for many SOCs with large sensor fleets. Keeping tabs on sensor traffic statistics and "health and welfare" is a daily job, even when using network taps.**

Finally, at the bottom, we see a sensor placed directly in-line between Alice's and Bob's networks (e.g., a NIPS). Most modern COTS NIDS/NIPS appliances offer a monitoring mode where the sensor can sit in-line and passively listen to traffic without any intentional interference. Less robust sensors may "fail closed" (e.g., if an error occurs or the device loses power) and disconnect Alice and Bob. However, most best-of-breed products take great care to avoid this problem.

In every case discussed, the network sensor must be physically near the network devices it monitors. This usually means that the network sensor is in the same rack, server room, or building as the monitored network segment. While it is certainly possible to send a copy of network traffic to a distant sensor using a remote SPAN, long-range optical connection, or even a WAN link, doing so can become very expensive. In essence, this usually means doubling the network throughput from the source network equipment to the location of the sensor. This almost never makes architectural or financial sense. Physically placing sensors adjacent to their monitored network segment is almost always the cheapest option; as a result, effective remote management is essential.

We summarize the pros and cons of approaches to traffic redirection in Table 19.

All these traffic redirection options have implications for how we prevent the network sensor from compromise or discovery. First, the monitoring port or ports should not have an IP address assigned to them. This will minimize the likelihood that it will talk back out on the network or bind services to the port. It follows, of course, that the monitoring port(s) should be used exclusively by the monitoring program(s) on the sensor and not be used for management or other services.

Table 19. Traffic Redirection Options

What	Pros	Cons
Hub	Cheap Easy to install Can attach as many monitoring devices as there are free ports	Sensor only sees what links at its same speed, not what links on other speeds, leading to gaps in visibility. Network management personnel may not be okay with non-enterprise-grade devices in critical routing points in the network. Sensors will miss packets due to collisions. Inherent inefficiency in hub's handling of traffic may lead to network bottlenecks. It is nearly impossible to find hubs that operate at gigabit (or above) speeds, limiting use to links that are throttled at 100 Mb/s.
SPAN	Free to use if monitoring points already have managed switches in place, which is very likely RX and TX are combined; one network cable off a SPAN port can plug right into a sensor. Straightforward for monitoring traffic from any device hanging off a switch (such as a firewall, WAN link, or cluster of servers) Can attach as many monitoring devices as there are free ports Can be used to monitor network core, such as spanning multiple ports off a core switch or router	An adversary with access to the enterprise network management platform (e.g., a terminal access controller access control system [TACACS] server) can disable monitoring feeds to the sensor. Some older or cheaper switches support only one SPAN port per switch, meaning additional switches may be needed if more than one sensor is desired. When spanning traffic from multiple source ports, the destination SPAN port may become oversaturated if the source ports' traffic aggregate bandwidth exceeds the SPAN port's speed. Changes to VLAN or port configurations after initial SPAN configuration can partially or completely blind the network sensor *without the SOC necessarily realizing it*.
Tap	Essentially invisible from a logical perspective. They only operate at layer 1, meaning the adversary does not have an obvious target to exploit or circumvent. Should not alter packets in any way Active network regen taps support multiple monitoring devices.	An additional device (albeit usually well built and simple) that can fail is introduced into critical network links. Only appropriate when observing conversation between two networked devices (as opposed to many with a network switch SPAN), as is often the case in perimeter network monitoring. Every monitoring point requires the purchase of an additional tap device. With a passive tap, RX and TX lines need to be recombined; some sensor technologies do have the internal logic to do so. Passive network taps only support one monitoring device.

Table 19. Traffic Redirection Options

What	Pros	Cons
In-line	Sensor can actively block traffic, depending on rule set.	If sensor goes down, it may cut off communication unless resiliency features are built in (e.g., "fail open"). Some sensor technologies may introduce packet latency or packet reordering, which in turn can sometimes degrade network quality of service or make the sensor detectable. More than one monitoring device means serial attachment of devices in-line, each being a separate point of failure.

In more extreme cases, this isn't enough. There are approaches we can take at a hardware level to ensure that the IDS's monitor ports cannot interact with the monitored network. In the case of a network tap, this has already been done. These devices, by their very nature, make copies of traffic that are destined for the sensor's monitoring port and do not accept response traffic.

In the case of COTS in-line devices, the opposite is the case. We must trust the implementation of the sensor technology to ensure the device will not interact with the network it is monitoring (except for blocking traffic, when appropriate). In the case of FOSS in-line devices, this duty falls upon both the authors of the FOSS monitoring software and the integrators of the FOSS sensor platform.

If we are using a switch to span traffic to a sensor, we have some interesting options at our disposal. We must prevent the sensor from transmitting packets back out on the network, thereby mitigating the effects of most attacks against the sensor (with the exception of DoS). However, the network switch that is sending the sensor packets must also establish a layer 1 link before it will transmit packets.

As a result, we can't just clip the transmit leads onto the Ethernet cable running from the sensor to the switch—the switch will think there is no device attached and drop the spanned traffic. In this case, we have two options. The first is to fabricate a receive-only Ethernet cable, as is specified in [240]. Compared to other options, this approach is quick, cheap, and easy.

If the passive network sensors are to reside on a network with a significantly different trust level than the network being monitored, a more robust solution such as a data diode may be necessary. Many commercial data diode solutions exist (e.g., those by Owl [241] or HP/Tenix [242]). However, these tend to be expensive and don't really meet our needs. We need an unaltered stream of network traffic data directly from the monitored network (something most commercial data diodes don't do). Also, in large enterprises, we may have many dozens or hundreds of sensors, necessitating an economical solution.

It is possible to use three copper to fibre network transceivers to do this, as described in [243] and shown in Figure 29.

In Figure 29, we see the three fibre transceivers in light green. Each transceiver has three connections: (1) fibre transmit (TX), (2) fibre receive (RX), and (3) a copper RJ45 combined receive/transmit connection (Cu). They are connected in such a way that data can only flow from the monitored network to the sensor. In this arrangement, we use the third data diode at the bottom of the diagram to fool the data diode on the top right that it has a valid Ethernet link. What is really happening is that only the RX link from the transceiver on the monitored network side is connected to the TX link on the transceiver connected to the network sensor, at the top left.

For further information on network sensor isolation techniques, see Chapter 3 of [9] and [240].

10.2 Designing the SOC Enclave

We've shown how to deliver network traffic to monitoring systems, but this is just the beginning. We must also cordon off SOC systems from the rest of the constituency but still allow them to intercommunicate and gather data.

Virtually every SOC in existence is responsible for defending a large number of IT assets. These assets are usually bound through a series of transitive trust relationships such as being members of a Windows domain. Most typically, the vast majority of Windows servers and user desktops are members of one or more domains that are

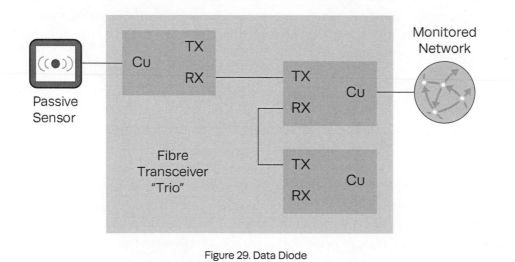

Figure 29. Data Diode

likely part of a forest. One of the most common goals for an attacker is to gain privileged domain rights, possibly through a domain administrator account. This allows an attacker to move laterally across any system that is a member of the domain or administered from the domain.

Any aspect of the SOC's mission operating on these domains presents a very serious risk. Consider a SOC that operates a fleet of IDS sensors and SIEM. It is fair to assume that the APT with domain administration privileges can install keyloggers and RATs on any system on the constituency domain. Even if SOC IDSes and SIEM systems are not joined to the domain, they may still be compromised if analysis and maintenance is performed from desktops that are. As a result, we arrive at a key recommendation:

> **Never join SOC monitoring infrastructure, sensors, analyst workstations, or any other SOC equipment to the general constituency's Windows domain.**

Many SOCs follow this rule: Any aspect of the SOC mission or its data that flows across systems that are joined to the constituency Windows domain should be considered compromised. The SOC is considered the "inner keep" of the constituency castle and should be the least likely asset to be compromised. As a result, the SOC must be even more vigilant in securing its systems against compromise. This often means shortened patch cycles and a robust security architecture.

Modern IT enterprises leverage centralized system administration and user authentication for a reason; so should the SOC. Many SOCs operate their own out-of-band network—the SOC enclave—with its own independent Windows domain. In heavy UNIX/ Linux environments, they may choose to leverage an equivalent (e.g., centralized LDAP, Kerberos, or Network Information Service [NIS]). This centralized user management can also be extended to the monitoring technologies (IDS, etc.) in other physically disparate locations. Taken to the extreme, some SOCs may actively choose *not* to join some domain authentication-enabled SOC systems to the SOC domain, as an extra measure against internal sabotage. Even if the SOC domain went down, would analysts still be able to log into IDS consoles? Having emergency use-only local user accounts with different user names and passwords as the Windows domain is always a good idea.

Typically, each analyst will have at least two desktop systems at hand: one workstation for CND monitoring and one standard enterprise desktop for email, Web browsing, and business functions. Where a SOC maintains watch over several disparate networks, analysts may have multiple workstations with a keyboard, video, and mouse (KVM) switch, or access through multilevel desktop systems or browse-down capabilities. Maintaining this separation introduces some inconvenience for the SOC analyst, but this is outweighed by maintaining the highest level of integrity. One way or another, the analyst must have access to three things: (1) monitoring tools, (2) constituency network(s), and (3) the Internet (per Appendix F).

We've established the need for an independent domain of trust for the SOC. How do we leverage monitoring tools at disparate locations while keeping them logically separate from the rest of the constituency? Before examining various options, here are some questions to consider:

What is the general cybersecurity posture of the rest of the constituency?

- Is malware running rampant?
- How well maintained are general user systems and networks? If the SOC depends on a general pool of IT resources and sysadmins, will they become a liability?
- What proportion of the user population has special system privileges, putting them in a position to gain unauthorized access to SOC systems that aren't strongly separated from the constituency?

How many different geographic locations will likely have to host monitoring technologies such as network sensors or log aggregators?

- If it's just one or two places, isolating monitoring equipment from the enterprise can be relatively simple.
- If the SOC has to put IDSes in 15 different countries, the approach will have to be highly scalable and more cost sensitive.

What kinds of logical separations are already in use and can they be trusted?

- What WAN technologies (MPLS, Dynamic Multipoint Virtual Private Network [DMVPN], ATM, VRF/GRE, etc.) that allow logical separation of assets based on trust domain are core competencies of the constituency? What are their relative strengths and weaknesses?
- How securely are the WAN and LAN segments administered? Is router administration done via in-band telnet with generic passwords, or is there an out-of-band management network used exclusively for router administration with SSH connectivity and TACACS running two-factor authentication?
- Or is the constituency run on a completely flat, Internet-connected network?
- Are the firewalls trustworthy enough to hang the SOC network off one leg without further protections?
- Is the SOC collocated with network operations so that there's increased trust and communication in their ability to manage SOC network infrastructure?

What equipment comprises most of the SOC's remote monitoring capabilities?
- FOSS monitoring technologies running on FOSS OSes (e.g., Linux or BSD) come with a number of native encryption capabilities, such as stunnel.
- If the SOC has a number of Windows systems or appliances remotely deployed, what options does the constituency network team offer to transport SOC data back in a protected manner?

What proportion of the SOC's monitoring assets must operate in band?
- Does the SOC plan to support a widespread deployment of HIPS or NIPS?
- Can the SOC rely mostly on passive NIDS sensors, with select placement of IPS devices?

How much money does the SOC have?
- Can it afford to purchase (and potentially manage) separate networking infrastructures such as switches, routers, and firewalls? If there are multiple ops floors, does it make sense to own and maintain encryption devices (e.g., VPN) between them? Can the SOC trust the network shop to administer these devices?
- If the SOC has to deal with multiple independent zones of trust, what are the costs associated with the placement of monitoring equipment and the SOC enclave? Can the SOC afford to maintain a presence on each?

In reality, this is not a terribly complex decision to make, and most SOCs follow a number of similar patterns. Well-resourced SOCs usually establish their own enclave that is firewalled off from the rest of the constituency. Ideally, they are able to leverage the following best practices when constructing their monitoring network:
- The SOC has its own Windows or Linux workstations used for general monitoring and analysis, usually bound together as a Windows domain that has no trust relationship with the rest of the constituency.
- Malware detonation, reverse engineering, forensics, or other high-risk activities are performed on isolated/stand-alone systems or virtualized environments.
- Assets local to the SOC (such as all ops floor systems) connect through network switches and other infrastructures that are completely separate from the constituency.
- The SOC's local systems are protected from the rest of the constituency by a modestly sized COTS or FOSS firewall.
- All passive network sensors are separated by receive-only cables hanging off switch SPANs or network taps.
- Local network sensors are connected directly to SOC network switches.

- Network sensors at sites with only a handful of SOC devices are managed through SSL/TLS–encrypted tunnels such as stunnel [244], software VPN, or vendor-provided encryption schemes.
- If there are several remote sites with a dozen or more pieces of monitoring equipment (IDSes, log collection servers, etc.), the SOC sometimes may choose to put them behind a hardware VPN concentrator. This establishes a VPN tunnel back to a concentrator near the SOC ops floor, providing a single tunnel for all management protocols and enhanced protection.
- Host instrumentation such as HIPS is managed by a central management server, potentially residing within the SOC enclave with a hole punched through the SOC firewall.
- Network traffic (PCAP), log, event, and case data reside on NASes or a modest SAN devoted to the SOC, not shared with the constituency.

It should be noted that even if SOC analysis systems are placed out of band, we shouldn't forget about systems that are frequently used for analyst-to-analyst collaboration (e.g., webcams, instant messaging chat, wikis, or, perhaps, even VoIP phones). Imagine, for instance, an APT that is listening in to the SOC's VoIP calls and learns that it has been detected. While this example may be far-fetched, it is certainly plausible and is something the SOC should consider when designing guidelines for how remote analysts collaborate.

One thing to note is that it may be tempting for the SOC to hang its equipment directly off general constituency network switches, with only VLAN separation. This is not recommended, for two reasons: (1) VLAN hopping [245] can circumvent this, and (2) accidental misconfiguration, compromise, or sabotage of these switches would open a backdoor to SOC systems, circumventing any kind of in-place firewall protection.

These best practices are depicted in Figure 30.

This architecture is optimal for medium to large SOCs that have a dedicated administration team. Smaller SOCs may have to take shortcuts such as comingling their assets on general constituency networks. Tiered and distributed SOCs can also leverage this architecture by adding combination VPN concentrators/firewall devices at remote sites where small teams of analysts reside. They also may choose to place log aggregation devices in band and then feed some or all of that log data to the SOC's out-of-band systems.

In protecting the SOC enclave, there are some additional controls that should be observed:

- SOC sysadmins maintain top-notch vigilance with patching SOC systems and updating sensor signature patches.
- The SOC's analysis environment is on a platform that is more resistant to malware infections than general constituency workstations. "Sandboxing" routine analysis

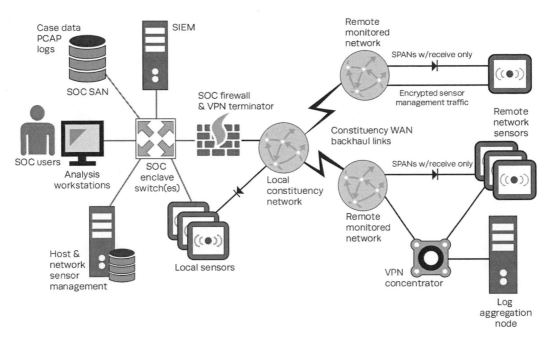

Figure 30. Protected SOC Architecture

functions through virtualization or on a non-Windows OS may help, understanding that usability and workflow must not be hindered.

- SOC sysadmins and analysts avoid use of shared user accounts, especially for privileged access, understanding the following:
  - The ultimate goal is to support attribution of privileged actions to a specific user in the unlikely event of configuration errors, compromise, or sabotage.
  - It is very hard to remember a different password for each disparate system.
  - Tying authentication to the domain for every type of device usually isn't possible.
  - Some general monitoring systems such as those projected on big screens on the ops floor will require generic accounts.
  - It may be best to strictly limit use of generic "root" or "administrator" accounts to emergency situations, whereas, normally, all sysadmins have their own account with administrator roles or rights.
- SOC equipment is under tight physical control.
  - The SOC has its own physical space with limited proximity badge access.
  - Local SOC equipment is in a SOC-controlled server room, server cage within the local server room, or at least in racks that close and lock with non-generic keys.

- There is robust (but not overzealous) logging for all SOC systems, such as sensor management servers, domain controller(s), and firewall(s).
- Both SOC sysadmins and a third party regularly review SOC logs for any evidence of external compromise, sabotage, or infection, thereby addressing the question, "Who watches the watchers?"
- The SOC may choose to use a different or more robust AV/anti-malware package than the rest of the constituency.
- Appropriate levels of redundancy are built into key systems such as sensor management servers, network switches, and log aggregation servers. The SOC may wish to establish a COOP capability as described in Section 4.3.5.

10.3 Sources and Methods

A SOC must interact with many other organizations. In order to gain their trust, a certain level of transparency is necessary. Giving the right people a sense of how a SOC gets the job done breeds respect and acceptance. However, there are some pieces of information that should not be shared with anyone outside the SOC without a compelling need.

For instance, were an adversary to obtain a list of monitoring point locations, it would then understand where there *isn't* coverage, allowing it to avoid detection. SOCs get requests from external stakeholders on a semi-regular basis for lists of their IDS tap points. Blanket requests for such information demonstrate a lack of sensitivity and awareness on the requestor's part to the realities of CND. The more people the SOC shares this information with, the more likely it is to end up posted on an Internet website or found on a compromised server.

In Table 20, we examine common pieces of information shared by the SOC and some likely circumstances under which it should and should not be released.

Table 20. Sharing Sensitive Information

What	Who Gets Access and Why
Monitoring architecture. High-level depiction of how the SOC monitors the constituency	Major IT and cybersecurity stakeholders such as constituency executives, security personnel, partner SOCs, and others listed in Table 25 (Appendix A)
Monitoring tap points. Exact locations of sensor taps and full details on how they're protected	No one outside the SOC except those who maintain or deploy sensors, if this function is separate
Monitoring hardware/ software versions, patch level	No one outside the SOC except those who maintain or deploy sensors, if this function is separate
Network maps	Organizations with a need to understand the shape/ nature of the consistency network, such as IT ops, network administration, or the offices of CIO/CISO
Vulnerability lists and patch levels (scan results)	Those who have purview over the vulnerability status of the constituency, including sysadmins, ISSMs, or CIO/CISO
SOC system and monitoring outages	Those directly above the SOC in its management reporting chain, such as the CISO or head of IT operations
Observables, indicators, and TTPs, including IDS signatures and SIEM content	External organizations (such as partner SOCs), with some potential exclusions for extra-sensitive signatures or insider threat indicators
Major incidents (possibly in progress)	Those directly above the SOC in its management reporting chain, possibly the CISO, in accordance with legal or statutory reporting requirements, such as with a national SOC
Incident details, including personally identifiable information	The appropriate investigative body, such as law enforcement or legal counsel
Incident roll-up metrics and lessons learned	Those directly above the SOC in its management reporting chain, possibly the CISO or CIO
SOC incident escalation CONOPS and flowchart	Any interested constituents
Audit logs for non-SOC assets	Individuals assigned the responsibility for monitoring IT asset audit records, such as sysadmins and security personnel
Raw security events (e.g., IDS alerts)	Under rare circumstances, select parties within the consistency (e.g., TAs) can help support CND monitoring because of their extensive knowledge of local systems and networks.

Ideally, all of this data should be located exclusively in the SOC's enclave. Many SOCs will host a website that may include a protected/authenticated portion where some sensitive documents will be posted, such as the incident escalation flowchart or network maps.

Considering the secrecy inherent in enterprise-class CND, the SOC is encouraged to provide some details of how it executes the CND mission to the constituency. Sharing information about the types of techniques used—without giving away the "secret sauce" on exactly how it's done—will go a long way toward building trust with interested parties. The SOC is advised to share some details with select constituents about its TTPs for spotting external adversaries. This presents a lower risk than sharing details about its insider threat program. Even high-level architecture diagrams are okay to share on a limited basis, so long as device details (e.g., IP addresses, host names, and software revisions) are removed. Moreover, when the SOC demonstrates forward-leaning, robust capabilities, it informs users that their actions are indeed being monitored. This may potentially ward off some miscreant activity. The key, though, is not disclosing so much that a malicious user knows how to circumvent monitoring.

Chapter 11

STRATEGY 9: Be a Sophisticated Consumer and Producer of Cyber Threat Intelligence

■ .... ■

The rise of the APT [246] renders traditional network defense techniques inadequate. Static methods and tools such as system patching, firewalls, and signature-based detection, by themselves, are not enough to defend against client-side attacks and custom-built malware. In order to level the playing field, the defender must orient his mind-set and capabilities to a dynamic, threat-based strategy. Analysts must consume, fuse, produce, and trade information about the adversary on an ongoing basis. This new trade in cyber threat intelligence has led to the creation of a new entity, the Cyber Threat Analysis Cell (CTAC), specially geared toward defending against the APT by maximizing the development and use of cyber threat intelligence across the cyber attack life cycle.[1]

In our ninth strategy, we discuss how SOCs may stand up a CTAC. It discusses the mission, resources, deliverables, and costs associated with CTAC operations. It also provides a roadmap for creating a CTAC, and references to resources with supporting information. Even if the SOC doesn't contain a CTAC, it may regularly consume, fuse, and redistribute cyber threat intelligence; we cover this topic as well.

---

1 Definition of the cyber attack life cycle, also known as the cyber kill chain, is consistent with [52].

11.1 What Is a Cyber Threat Analysis Cell?

A CTAC is a team composed of advanced security analysts organized to detect, deny, disrupt, degrade, and deceive the APT.[1] Its existence presupposes the existence of a routine cybersecurity monitoring and incident response capability, such as a SOC. A CTAC depends on the capabilities of—and is part of—a SOC.

Operating the CTAC enables the SOC to be a sophisticated consumer and producer of cyber threat intelligence (often shortened to "cyber intel"). While addressing the APT is the primary interest of a CTAC, its TTPs enhance all aspects of a SOC's capabilities. A designated group within the SOC may be considered a CTAC if it routinely performs all five of the following core functions:

- **Extraction of indicators of compromise**, through a combination of digital artifact examination, static code analysis and reverse engineering, runtime malware execution, and simulation techniques
- Routine **ingest of cyber threat intelligence** reporting and news from a variety of sources
- **Fusion** of locally derived and externally sourced cyber threat intelligence into signatures, techniques, and analytics intended to detect and track the APT
- Active participation in **cyber intelligence threat-sharing groups**, typically composed of other SOCs in a similar geographic region, similar supported organizations and industries, or both
- **Advanced incident analysis and response support**, such as digital forensics of memory and hard drive images.

The CTAC often is composed of some of the SOC's most experienced analysts. The rapidly changing nature of APT TTPs often pushes a CTAC to perform the following additional functions:

- Creating and tuning **advanced analytics** to detect complex or advanced attack patterns, such as those used to detect and track the APT in the SOC's SIEM tool
- Developing focused, finely scoped **custom tools** that better enable the CTAC to detect, observe, contain, or block the APT at different stages of the cyber attack life cycle
- Operating and populating a **threat knowledge management capability**, allowing SOC analysts to connect disparate but related adversary activity, incidents, indicators, and artifacts
- **Trending** and **reporting** on activity and incidents attributed to the APT

---

1   This definition follows that of a CSIRT in [42] and [43], and courses of action mentioned in [52].

- **Hunting** for the presence of the APT on monitored networks, such as through targeted monitoring efforts or "sweeping" for indicators that strongly correlate with APT activities
- **Honeypotting** and other methods that allow the CTAC to observe the adversary *in situ.*

While the CTAC must maintain keen awareness of myriad threats that reside all over the Internet, its response actions are defensive in nature and are limited to the scope of systems it is asked to defend. Offensive "hack back" type actions are outside the scope of what the CTAC performs. Where appropriate, the CTAC should cultivate a good working relationship with entities empowered to perform cyber investigations and potentially direct responses against adversaries, such as some types of law enforcement.

11.2 What Does It Provide?

Operating a CTAC provides a number of primary, first-order benefits, and many more second-order impacts that enhance cybersecurity for the entity it serves. They are as follows:

11.2.1 Primary Impacts

- Higher confidence in the efficacy and completeness of incident response actions
- Decreased proportion of APT attacks that are successful
- Decreased time the adversary is able to maintain presence *without being detected*
- Deeper threat awareness through direct knowledge of adversary TTPs throughout the cyber attack life cycle
- Enhanced SA through more informative and more thorough threat and incident reporting
- Increased context and link between incident activity and mission impact.

11.2.2 Secondary Impacts

- Focused, higher impact use of cybersecurity resources (time, funding, talent, organizational political capital, and will) on threat-focused prevention, sensoring, analytics, and response capabilities
- Simpler, faster SOC service delivery through reduced reliance on external parties that perform malware and forensic analysis
- Improved morale, stemming from sense of fraternity and fellowship with partner SOCs
- Better attraction and retention of SOC personnel through expanded career advancement and membership in a world-class capability
- Increased differentiated value in contrast to other cybersecurity stakeholders

- Greater value and return on investment of all SOC service offerings through "trickle down" of CTAC expertise and lessons learned to other areas of cybersecurity operations
- Increased stakeholder responsiveness due to SOC's ability to articulate meaning of adversary activities in context of mission, and confidence in efficacy of response actions
- Substantial savings of effort by leveraging solutions and cyber threat intelligence from partner SOCs
- Enhanced awareness of organization's threat profile and likely targets of adversary attack
- Increased insight into gaps in SA and complementary motivation to fill those gaps.

11.2.3 Potential Work Products

Each CTAC produces a set of deliverables and artifacts on a routine basis. Some of these deliverables are easily recognized briefings or papers, whereas others take the form of inputs to an online knowledge base or updates to tools or technologies used by the CTAC.

Table 21 lists some of the written artifacts a CTAC is likely to produce. Any of these work products that are meant for external consumption constitute the cyber intel that the CTAC produces. **Cyber intel** is defined as formal or informal information and reports from SOCs, CTACs, commercial vendors, independent security researchers, and independent security research groups that discuss information about attempted or successful intrusion activity, threats, vulnerabilities, or adversary TTPs, often including specific attack indicators.

In addition to traditional, written artifacts, the CTAC is likely to apply substantial efforts toward non-traditional tangibles that are also worthy of recognition, such as those in Table 22.

11.3 How Does the CTAC Integrate into IT and Security Ops?

In order for the CTAC to be successful, it must work hand in hand with every part of the SOC, and with a number of stakeholders outside its parent SOC. The CTAC, by itself, has very few tools to detect and track the APT, and even fewer to respond to the APT. It is heavily dependent upon the sensoring and analytical tools furnished by the SOC, and the blocking/response capabilities and responsibilities vested in other areas of IT operations.

Table 21. CTAC Artifacts

What	Discussion	Typical Frequency
Case tracker notes and reporting	Incidents that are targeted in nature or are related to a known APT may be referred to the CTAC for in-depth analysis. The CTAC's working notes, activities, recommended follow-up, and other analyst-to-analyst communications are recorded in the SOC's incident case tracking capability.	Daily–Weekly
Formal incident write-ups	Particularly notable incidents handled by the CTAC may deserve formal documentation or presentation outside the scope of what is captured in the SOC's incident tracking capability. This may take the form of presentations, written reports, or sometimes both, authored by the CTAC or co-authored by the CTAC and SOC incident responders.	Monthly–Quarterly
Cyber threat tipper	Short, timely information "tipped" to a cyber threat sharing group within minutes or hours of identifying activity as likely relating to a targeted intrusion attempt. The information may be as simple as the sending email address, subject, and attachment names for a spear phish or URLs for a drive-by download, or it may include preliminary malware analysis, such as callback IP addresses, domains, file hashes, persistence mechanisms, and sample beacon traffic.	Daily–Weekly
Short-form malware report	Two- to four-page report that provides some indicators and information regarding an observed piece of malware. Usually stems from malware that took one or two days of static or dynamic analysis to understand.	Weekly–Monthly
Long-form malware report	Three-plus-page report that provides detailed indicators and reporting on an observed piece of malware. Generally stems from a deep-dive reverse engineering effort that took several days or weeks to accomplish. Typically includes a full description of the malware sample's functionality, any encryption used, and its network protocols used for command and control. It may include additional tools and techniques developed alongside the analysis, such as malware network protocol decoders, and ways to unpack and extract encryption keys and other indicators from malware samples within the same family.	Monthly–Quarterly
Adversary and campaign reports and presentations	Briefs that discuss the TTPs, intent, activity seen, incidents, etc., stemming from a named adversary or adversary campaign. Usually strings together activity seen from multiple incidents and/or several months of reporting.	Monthly–Quarterly

11.3.1 Inside the SOC

For each part of the SOC, the CTAC has a responsibility to provide timely SA regarding the TTPs, activities, and impacts of adversary activity. In exchange, each section of the SOC must enable the CTAC's mission in different ways, as detailed in Table 23.

Table 22. Other CTAC Work Products

What	Discussion	Update Frequency
Malware analysis and forensics environment	In order to understand the nature of suspected malware, the CTAC will require an environment in which to perform static decomposition/disassembly and runtime execution/simulation of malware. This environment is solely used by the CTAC, requires a set of very specialized software, and usually is operated on a set of hosts or a small network that is well isolated from all other computing resources. As a result, the CTAC must be responsible for maintaining and updating the capabilities of this malware analysis environment.	Monthly–Quarterly
Threat knowledge management tool	The CTAC will require a means to track and link adversary activity, campaigns, indicators, events, associated malware samples, and associated PCAP samples over time. This capability stands apart from the SOC's incident case management system, but may support some integration with it. The CTAC is the primary author and curator of content in this database; it may be used and referenced by all other analysts in the SOC.	Daily–Weekly
Indicator lists	Part of the CTAC's job is to aggregate various indicators of compromise (suspicious IP addresses, domains, email addresses, etc.) from external cyber intel reporting and its own malware reverse engineering. These indicator lists are primarily used to generate signatures and other detection content in the SOC's tool set (NIDS, SIEM, HIDS, etc.). They may also be housed inside—and generated from—the CTAC's threat knowledge management tool.	Daily–Monthly
Sensing and analytics enhancements	Administration and tuning of sensors and analytic systems, such as IDS and SIEM, are usually not the responsibility of the CTAC, but of sensor O&M within the SOC. However, it is sometimes most efficient for the CTAC to directly translate knowledge of the adversary into signatures or use cases. In such cases, the signature's author in CTAC will likely work with sensor admins to document and operationalize it.	Weekly–Quarterly
Custom tools or scripts	The CTAC will notice gaps in its capabilities that cannot be satisfied through FOSS or COTS solutions. This is especially the case as the CTAC matures over time and has to deal with unique or particularly advanced threats. Quarantining and observing the adversary, parsing or simulating command and control traffic, or ingesting foreign sources of data into a tool are three examples where custom code may be needed. Custom tools are spun off on an irregular basis, usually developed very quickly, and don't always reach full maturity before they are no longer needed.	Irregularly; Monthly–Quarterly

Table 23. CTAC Relationship to SOC Elements

SOC Element	Role
SOC Chief	Top cover for any operations and initiatives that require it; upward reporting of important products and successes
Tier 1	Identification of activity that might be related to actors of interest to the CTAC, based off of CTAC-developed SOPs and CTAC-developed sensor rules, SIEM configurations, and similar technology; escalation of incidents requiring expertise or capabilities of the CTAC
Tier 2	Heavy lifting for all incident response activities; escalation of incidents requiring expertise or capabilities of the CTAC
SOC engineering	Budgeting, acquisition, engineering, integration, and deployment of SOC tools and the SOC network enclave, including those that enable the CTAC mission, such as SOC workstations, network sensors, host sensors, and SIEM
SOC tools O&M	Day-to-day management and sustainment of SOC tools and the SOC network enclave, including those that enable the CTAC mission

In contrast to SOC Tier 1 and Tier 2, the CTAC is not generally responsible for the "daily grind" of ticket handling and closure. When the CTAC is handed a case, it may very well stay open for weeks or months while analysts attempt to understand the malicious or anomalous activity of concern. Unlike Tier 2, the CTAC's activities do not necessarily stem from a specific incident or alert. Members of the CTAC are expected to be self-starters, proactively looking for new ways to detect the APT across various stages of the cyber attack life cycle.

The CTAC must work with the rest of the SOC on developing and continually refining SOPs, especially regarding how the SOC's tiered analysis personnel triage and process activity detected by CTAC-developed means. The CTAC is also responsible for ensuring that SOC analysts have sufficient contextual knowledge to understand and take corrective action. Finally, the CTAC should continually solicit feedback on the performance of CTAC-developed processes, tools, and detection capabilities, to reduce the risk of wasted resources on false positives.

An agile defense must be mounted against the APT. Tools must correctly meet operators' needs, and go from concept to deployment in short order. To achieve this, it is necessary to bring the budget and personnel supporting SOC tool engineering and development directly into the SOC organization, consistent with the DevOps model [247]. This also allows the CTAC to become directly involved in tactical tool development and integration in a much cleaner, less politically contentious manner than if SOC tool engineering were located outside the SOC. Operating passive monitoring tools (such as IDSes) in a protected enclave will also allow the SOC to pursue the DevOps model with greater freedom.

The SOC will have to carefully balance the needs for agility and responsiveness with those for security, stability, reliability, and maintainability. SOC leadership should put appropriate controls in place to prevent the SOC's in-line prevention tools from unnecessarily impacting availability. However, in some cases, particularly for SOCs within government or heavily regulated industries, existing security and IT policies may come into conflict with the rapid development and deployment of custom-built and other non-traditional tools. This may require the SOC's champions to fight for a highly responsive and flexible interpretation of traditional policy-oriented security and IT processes.

11.3.2 External to the SOC

In most cases, the CTAC's relationship with parties outside the SOC should be indistinguishable from that of its parent SOC(s). For instance, most users and the IT help desk probably don't need to know that a CTAC exists; they just need to know that potential cybersecurity incidents should be referred to the SOC. Other parties, however, may recognize and interface with the CTAC directly, due to its special role in operations. These relationships are depicted in Figure 31 and discussed below.

The CTAC's intimate knowledge of the adversary will likely be of specific interest to IT and security executives, such as the parent organization's CIO, CISO, and CSO. The SOC can probably expect that the CTAC will provide monthly or quarterly threat briefings to interested executives. Providing these briefs is important, even if they're just informational: it builds trust in the SOC and helps justify its budget. If the SOC's parent organization has any parties that must maintain strong awareness of cyber threats, such as an industrial counterespionage or counterintelligence shop, the CTAC should consider collaborating with those groups as well. The CTAC will also require direct liaison authority with IT and security personnel throughout the organization, particularly as it relates to incident response.

When it comes to sharing cyber intel with other SOCs, the CTAC takes the lead. In some cases, this can be pair-wise sharing with one other SOC. However, nearly all CTACs participate in cyber threat intel sharing groups. These groups usually consist of a handful to several dozen other SOCs with some common attribute—usually geographic region, nationality, or business function such as government, industry sector, or education. In all cases, these relationships are almost always heavily reputation based, brokered at the analyst-to-analyst level. There must be a mutual sense that each participating SOC has something to add and that indicators will be protected; standing up a CTAC is the best way to gain substantive entry to such sharing groups.

The CTAC should also cultivate a relationship with relevant law enforcement organizations empowered to investigate cyber crime. The SOC and its champions must ensure that

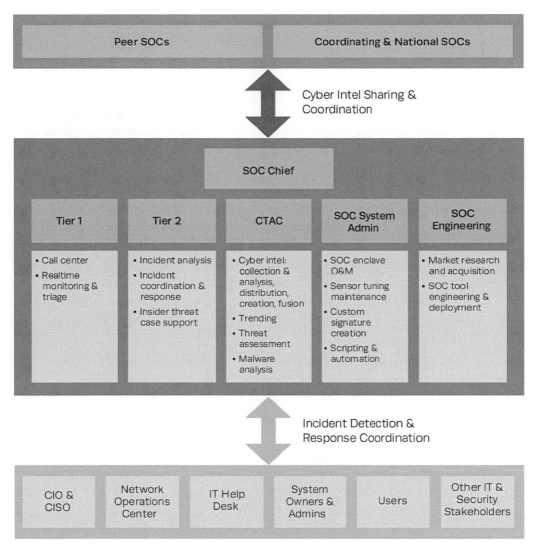

Figure 31. CTAC and SOC Relationships to Other Organizations

the CTAC has direct liaison authority with these outside parties, with broad but clearly defined authority regarding collaboration and disclosure.

During focused adversary engagements, it is the CTAC's goal to gain intimate knowledge of the adversary in the context of the impacted systems, mission, data, and users. This usually requires ad hoc instrumentation of enclaves and hosts at the edge of the network, and potentially redirection of adversary activity. As a result, the CTAC will have to maintain a close working relationship with system owners, sysadmins, and network operations for the duration of the engagement. Once again, this will also mean bringing in security and counterintelligence specialists, if they exist.

11.4 What Are the Prerequisites for Standing Up a CTAC?

Before committing resources to creation of a CTAC, SOC managers must first assess their organization's readiness for CTAC operations. In this section, we propose four questions that should help an organization recognize if the time is right to stand up a CTAC. If the answer to most or all of these four questions is a resounding "yes," then the time is right. If not, then it may be a good idea to focus on more foundational IT operations and SOC capabilities before initiating a CTAC.

1. Does the organization served have an in-sourced incident detection and response capability that:
   a. Has a defined set of users, sites, IT assets, networks, and organization(s) that it serves, known as its "constituency"
   b. Collects information on cybersecurity incidents from users and various sensors and data feeds, performs in-depth analysis, and in turn performs various response actions
   c. Documents and tracks incidents using a COTS, FOSS, or custom incident tracking capability
   d. Has separated incident monitoring, analysis, and response into two or more groups or "tiers" of increasing focus and capability?

2. Is the SOC empowered through appropriate authorities, procedures, and executive support to:
   a. Refrain from blocking adversary activity when necessary to understand the full scope of an intrusion and adversary techniques, tactics, procedures, and intent
   b. Directly control allocation of budget for SOC tools and personnel
   c. Rapidly acquire, budget, install, update, and tune a wide range of commercial, open source, freeware, and custom-developed tools across the constituency in response to emerging threats and advances in defender and adversary technology
   d. Direct network operators to perform tactical blocking and redirection operations, such as firewall or proxy rule changes, DNS blackholing, logically isolating an end host, and quarantining malicious emails or email attachments
   e. Operate all of its non-in-line capabilities, such as passive network monitoring, SIEM, and incident case handling, in an out-of-band protected SOC enclave

    f.   Collect, retain, and search various incident-related information such as full-session PCAP data, malware samples, and hard drives from compromised systems?

3.   Can the SOC answer the following questions with respect to its constituency?

    a. What sources of data are of greatest value in finding initial indicators of malicious or anomalous activity

    b. What sources of data are of greatest value in running suspected malicious or anomalous activity to ground

    c. How many incidents are related to phishing, pharming, and direct attacks each month

    d. What is the meaning and disposition (true or false positive) for a random sampling of events from each major security-relevant data feed and sensor technology (NIDS, host-based AV, proxy AV, etc.) received by the SOC

    e. What are the major limitations in its visibility?

4.   Is the SOC's parent organization willing and able to expand the resourcing to its SOC, in an effort to answer the following questions?

    a. What organizations are behind suspected cyber intrusions and what is their intent

    b. How is the adversary activity seen by the SOC related to adversary activity observed by other organizations in similar lines of business or geographic region

    c. How can the SOC increase confidence that its steps taken toward incident response ultimately result in deterring or denying adversaries' ability to achieve their objectives

    d. What is the impact of incidents to major lines of business or mission function?

11.5 What Do We Need and How Much Will It Cost?

Assuming that the SOC has fulfilled the prerequisites for standing up a CTAC, most of the costs associated with operating a CTAC should already be recognized and budgeted. In this section, we break down the investments needed to add a CTAC capability to a SOC.

11.5.1 Foundational Capabilities

For SOCs ready to stand up a CTAC, most large tool investments needed by a CTAC should already be in place; some may need to be extended or expanded. They are as follows:

- Network IDS sensors capable of custom signature support, and full traffic capture (PCAP) in key places like Internet gateways, WAN uplinks, and firewall boundaries (See Section 8.2.)

- A well-tuned SIEM tool that includes advanced real-time correlation and the ability to craft custom analytics (See Section 8.4.)
- Direct access to a variety of security-relevant data feeds, fed into SIEM or some other enterprise-grade log management solution with retention of the most recent 90 to 180 days of log data, preferably longer (See Section 9.2.)
- Host protection and sensor suites that include AV capabilities (See Section 8.2.9.)
- An out-of-band SOC enclave network that has no trust relationships with outside networks, such that SOC analysis systems and SOC data are protected, even if the APT has an active presence on monitored networks. (See Section 10.2.)

After the standup of the CTAC, the SOC can expect that the operating and upgrade costs of these tools may remain the same or go up. The CTAC will likely identify gaps in tool functionality and new opportunities for enhancements that directly support its objectives, but will also directly benefit other parts of the SOC. Investment focus areas will likely include expanded sensor coverage, advanced analytics, and longer log and PCAP data retention. Examining existing tool capability gaps will help the SOC estimate the costs to meet the CTAC's needs, even in advance of its standup. Here are some tips to consider when justifying these expenditures:

- Enhanced capabilities requested were most likely necessary to fundamental SOC functions, even in absence of the CTAC; the formation of the CTAC just made their absence more acute.
- Some capabilities, such as robust log generation and collection, are part of IT best practices or regulations that the constituency should already be following.
- In cases where more of a given capability is needed, such as longer data retention or expanded sensor coverage, economies of scale, in combination with scheduled recapitalization, could be used to drive down any perceived change in cost.
- In resource-constrained environments, some SOCs may be driven to compile metrics in building a business case for certain capabilities, which can be very time-consuming; if at all possible, the easiest way to overcome this challenge is to leverage an executive "champion" for SOC causes.
- Consider investments that have a very shallow cost increase when scaling out a solution, such as network sensors that leverage FOSS software and commodity server hardware.

The CTAC's thirst for robust logging and monitoring may trigger second-order impacts on systems belonging to other IT stakeholders. The CTAC will likely need high-volume data feeds from firewalls, email gateways, Web proxies, Windows Domain Controller/Active Directory, and DNS systems. If these systems are antiquated or nearing their performance limits, the addition of robust logging may enhance justification for a much-needed upgrade, even though commonsense logging practices don't necessarily cause a substantial increase in system load. These impacts should be recognized and accounted for in advance, even though the SOC is usually not the appropriate party to bear any resulting financial burden. This is especially important if such systems are provided through outsourced, performance, or service-based contracting, meaning enhanced logging may require contract modification or transfer of additional funds.

11.5.2 Analysis Environment

One of the most important new tool investments needed to operate a CTAC is the hardware and software used to support malware analysis and digital media forensics. While the SOC may choose to process the bulk of its digital forensic artifacts on its existing enclave network, the CTAC should conduct malware analysis on robust workstations disconnected (e.g., "air gapped") from other enterprise networks. This is done primarily to preclude the inadvertent spread of malware to other SOC or enterprise systems. That said, it is increasingly important that runtime analysis systems can also be connected to the Internet through a "dirty" connection not attributed or connected to the main corporate network. This is necessary in situations where gathering second-stage malware is desired. Being able to easily enable and disable Internet connectivity from runtime analysis systems is a good design goal for the analysis environment.

The CTAC's runtime malware analysis capability most prominently features a series of well-instrumented victim platforms hosted inside a virtualization environment. This is often achieved using a commercial or open source malware sandboxing and analysis platform, such as FireEye AX [248], ThreatTrack ThreatAnalyzer [80], or Cuckoo Sandbox [249].

Automated sandboxing tools will only take the analyst so far. Manual runtime analysis is sometimes needed, such as in cases where human interaction is required to trigger malware behavior, malware sleeps for a long time, malware will not run in a virtualized environment, or malware does not consistently execute. Also, analysts must often simulate both the attacker's command and control servers as well as the victim systems, meaning additional hardware and software investments are needed. For instance, gleaning the beaconing addresses used by a first-stage piece of malware is relatively straightforward, given a single copy of VMware Workstation [250], Microsoft Windows [251], Sysinternals tools [252], and Inetsim [253]. Fully understanding a RAT's command and control is more involved, though there are some frameworks such as ChopShop that can help [254].

The CTAC's static malware analysis capability should include myriad tools used to unpack, disassemble, decompile, trace, and analyze malware samples. Quintessential tools for this are IDA Pro [255], WinDbg [256], and OllyDbg [257]. These are just the start; various other tools, utilities, and scripts should support:

- Windows Portable Executable (PE) header analysis
- Executable unpacking
- AV scanning (utilizing multiple engines to reduce false negatives)
- Document and metadata analysis tools covering formats such as PDF, RTF, and Microsoft Office
- Scripting and runtime analysis tools covering languages such as Java, JavaScript, and Flash.

It is usually best to integrate all of these tools into one environment, as CTAC analysts will usually need to leverage a blend of these capabilities in order to accomplish their objectives most efficiently. Analysts may wish to have multiple virtual instances of their analysis environment available, with the ability to revert their environment to a known-good state, as is typically the case with runtime analysis.

The CTAC's digital media forensics capability must include tools that are capable of reading, storing, analyzing, and displaying the contents of digital artifacts, most notably hard drives and memory dumps pulled from systems involved in an incident. A software package capable of reading in the contents of a hard drive, analyzing, and displaying its contents is usually the focus of digital forensics work. Encase [258] and FTK [259] are examples of commonly used commercial forensics tools, though there are FOSS packages such as Sleuthkit/Autopsy [260] and SANS Sift Workstation [261] that offer compelling features.

Similarly, Volatility is a well-known tool used for memory forensics work [262]. The CTAC will likely seek additional tools to fill in for specific needs, ranging from reconstructing RAID arrays to ad hoc log analysis. The CTAC should also seek legal counsel in order to determine what tools and procedures are needed to perform legally sound handling of digital evidence to support potential future use in criminal, civil, or administrative proceedings. For most CTACs, this includes high-grade safes, evidence bags, and write blockers, such as those made by Tableau [263].

Some SOCs will purchase purpose-built forensics workstations such as the FRED [264] that have write blockers and mass storage built into them. It is sometimes helpful for SOCs to have portable digital evidence collection and analysis "flyaway" kits that can be taken to remote locations. This is especially the case when the team may need to travel to gather log data or memory images. In such cases, high-end laptops can be used, which will run software similar to the SOC's stationary forensics systems "back home." Like their stationary counterparts, these laptops will require enhanced processing power, memory, and disk capacity.

The treatment of tools and techniques involved in digital media forensics and malware analysis herein is only cursory. In order to fully scope the capabilities needed for this work, readers may consult various references on the topic, including those found here: [24], [25], [26], [27], [28], [29], [36], [37], [38], [39], [40], [265], [11]. That said, budgeting for most of these capabilities is relatively straightforward.

To summarize, CTACs should account for the following in their budget:

- Workstations capable of performing digital forensics, which may or may not double as their SOC workstations, including write blockers and expanded storage
- Commercial-grade digital forensics packages, as mentioned
- Workstations capable of performing static and dynamic malware analysis, including memory, processor, and storage resources necessary to run multiple concurrent virtual machines
- Commercial-grade virtualization software, and runtime malware analysis and sandboxing platforms, as mentioned
- Static malware analysis tools, as mentioned
- Operating system and software licenses sufficient to simulate computing environments targeted by commonly encountered malware strains
- Additional specialized commercial software tools that fill in for specific needs or tasks not covered above
- Storage for long-term retention of malware and digital forensic artifacts
- Modest networking infrastructure (network switch[es], network cable) to connect systems in the air-gapped malware analysis enclave
- Safe transfer mechanisms, such as removable storage or controlled network interfaces, that allow controlled transfer of malware samples and analytical findings into and out of the forensic and malware analysis environments.

To conclude our discussion of the analysis environment, here are some key tips to keep in mind:

- While tools like IDA Pro, Encase, and FTK are most often mentioned, and are some of the most expensive, they're only a start. A mature CTAC's tool set may weigh in at more than a hundred disparate software packages, utilities, and scripts.
- In order to meet its analysis needs, the CTAC will require new tools or tool updates quite frequently, often monthly or even daily. Some of these tools, such as those aiding in certain aspects of malware detection and analysis, may originate from less than reputable sources, or may be released under a FOSS license. Finding the right governance structure to manage risks and to minimize delays in updating tools will be important.

- It is very likely that some of these capabilities existed inside the SOC, at least in nascent form, prior to standup of the CTAC. With formation of the CTAC, their long-term funding and updates should be more formalized and sustainable.
- Specific tool choices are very task-driven, and often a matter of personal experience and preference. The only people in the enterprise who are qualified to select, engineer, or operate these tools are in the SOC. As a result, it is often best to defer build-out of the CTAC analysis environment until at least a few members of the CTAC have joined the SOC.

11.5.3 Threat Knowledge Management

As we have discussed, the CTAC differentiates itself from other cybersecurity entities by its deep knowledge of the cyber threat. It must build up a knowledge base about the adversary over time that all members of the SOC, new and old, can directly leverage. This knowledge base should:

- Allow analysts to draw connections between related items, allowing them to cross-reference or "pivot" among them
- Be organized, such that old information and artifacts can be retrieved quickly
- Be scalable, allowing the SOC to build up mass amounts of information and related contextual data over time, tracking dozens to hundreds of adversaries, hundreds of thousands of indicators, and TBs of digital artifacts
- Support continual pruning of stale or inaccurate information
- Allow analysts to easily refer back to original reporting for context and understanding
- Mark information by source and sensitivity, thereby supporting any necessary constraints on export and sharing
- Be carefully curated by members of the CTAC, and only include information that relates to known or suspected APT activity.

Different SOCs use different approaches to satisfy these needs, depending on the amount of data they wish to store and the number of analysts who will use it. Before it stands up a CTAC, it is very likely that the SOC will store digital artifacts on a SAN or NAS within the SOC enclave. In addition, the SOC may store information on particular adversaries in a Wiki [266], in its SIEM tool, or in some other tool, although it is likely that SOC analysts could use more time to fully populate it. Finally, if the SOC is already collecting indicators, there is a good chance it is housing them in a spreadsheet, a simple custom-built database, a SIEM, or some combination thereof.

These techniques may work for the CTAC when getting started, but they don't scale over time. Tools such as Palantir [267] support the node-link analysis, allowing analysts to "connect the dots" among related adversary activity. There are other capabilities that also support compilation and sharing of indicators, such as FS-ISAC's Avalanche [268], REN-ISAC's Collective Intelligence Framework [269], Lockheed Martin's Palisade™ [270], and Cyber Squared's ThreatConnect [271]. However, as of this book's writing, a leading set of commercial tools that meet the full scope of the CTAC's needs of threat knowledge management are still emerging. For this reason, MITRE developed an operational prototype called Collaborative Research into Threats (CRITs) [272], with the objective of satisfying all of the knowledge management needs stated above.

The cost to deploy a threat knowledge management tool is very tool-specific. It is very likely there will be a hardware cost involved for the storage needed to house all the digital artifacts, and modest server processing power to support concurrent use among SOC analysts. For COTS software, there is not yet a dominant licensing model, so no general costing model can be given. Additionally, there is an ongoing labor cost associated with importing and maintaining the knowledge base within the tool. That said, this is a core activity for the CTAC, and is usually less far less expensive than manually organizing indicators and manually querying sensor data for matches against indicators.

11.5.4 Remote Incident Response

Remote incident response and indicator sweeping tools are a critical enabler for hunting for the presence of the adversary on constituency systems and networks. They allow the SOC to perform such tasks as remotely imaging a system's memory or searching for a Windows Registry setting associated with a piece of malware. This capability can be a huge boon to the CTAC, both in terms of compressing response time and performing tasks across thousands of hosts simultaneously. These benefits often come at a price when widely deployed; their acquisition and maintenance costs often approach those of the SOC's five other foundational capabilities mentioned above.

These tools, such as HBGary Responder Pro [273], Mandiant Intelligent Response [201], and Encase Enterprise [274], are generally meant for widespread deployment across enterprise desktops and servers. They are also tightly integrated with their target platform, meaning they will require substantial integration testing and debugging prior to widespread deployment or upgrades. In both of these regards, the hurdles and cost associated with deploying these tools are quite similar to those of host protection suites. As a result, some SOCs may examine and prioritize costs associated with the various host-based protections at their disposal, especially with respect to scale and sustainability. Additionally, the SOC should determine whether existing IT management tools may already support some limited indicator sweeping capabilities, such as searching for suspicious file hashes.

11.5.5 People

One of the biggest challenges to standing up and operating a CTAC is finding and retaining qualified staff. This is already a challenge for anyone managing a SOC, but acutely so for the CTAC. A CTAC may be composed of two or three analysts for a small 10- to 20-person SOC, or for a CTAC that is just getting started. On the other hand, a SOC of 50 to 75 analysts may allocate ten analysts or more just to the CTAC. Keeping all of those positions filled can be challenging in a fiercely competitive labor market, especially for SOCs that require extensive background checks and clearance requirements, such as those in government and some areas of industry, or for SOCs requiring work in high-cost geographic areas.

CTAC work falls into roughly three or four areas of specialization, which we can use to frame the qualifications beyond those of general SOC analysts:

- Cyber threat intel collection, fusion, and sharing; participation in sharing groups: These analysts should have the same sorts of qualification and background as a typical Tier 1 or Tier 2 SOC analyst, including oral and written communication skills, analytical ability, and the ability to maintain a stream of productivity without frequent oversight. An analyst operating in this role for more than three to six months should be able to demonstrate substantial knowledge of adversary TTPs. Analysts who actively participate in threat-sharing groups should generally have at least 12 to 18 months of SOC experience under their belt.
- Digital forensics (media images): These analysts should have at least one to two years of experience in SOC work, and the traits of someone who is ready for Tier 2 analysis. Prior use and training on forensics tools is a plus, though this can be conferred through several weeks of vendor training.
- Digital forensics (memory images): In contrast to media forensics, analyzing memory dumps is typically far more complicated, and involves less straightforward point-click automation. Analysts performing memory forensics must have substantial formal education in computer architecture and programming. This narrows candidates almost exclusively to those with a degree in electrical engineering, computer engineering, CS, or computer forensics from an accredited college or university. In order to become proficient on forensics itself, candidates usually will have to undergo a combination of substantial self-study and formal training on forensics.
- Malware analysis and indicator extraction: Analysts should have background and training similar to those who perform analysis of memory dumps, with additional proficiency with compilers and machine assembly. Those without formal training on computer architecture or operating system design can perform limited runtime extraction of malware indicators such as beaconing, but will not be able to get very far into exploitation, static reverse engineering, or malware behavior.

There is, of course, substantial overlap among these specialties. It is encouraged for someone with a formal degree and prior SOC experience to rotate between malware analysis and cyber intel fusion roles. Most CTAC analysts with knowledge of the SOC's detection tools should be able to craft signatures for those tools based on cyber intel they handle. In a similar vein, most CTAC analysts should be able to support custom tool development, using whatever languages they are familiar with appropriate to the task. Pursuing this approach allows the CTAC to quickly surge in one area or another to meet changing ops demands.

The head of the CTAC should have experience in one of these CTAC roles, though this is not always possible due to the exploding demand for seasoned analysts. Having a background in some areas of law enforcement, intelligence analysis, or other parts of the SOC may translate well, so long as that person is capable of becoming intimately knowledgeable about adversary TTPs. In addition, the CTAC lead should be able to translate very technical concepts for non-technical audiences—critical when convincing executives that half of their business unit is impacted by a given incident, and why it's in their best interests to cooperate with the SOC. Level-headed, critical decision making must also be emphasized—knowing when to watch and when to block. The CTAC lead should be able to focus on these tactical issues, freeing up the SOC lead to focus on ensuring overall delivery of SOC capabilities.

Salaries for CTAC analysts generally range from that of a mid-level Tier 1 analyst beyond that of the most senior Tier 2 analysts. In the United States, it is very easy for a good malware analyst to find a job that pays well into the six-figure range, especially in major metropolitan areas and for SOCs that require extensive background checks or clearances. As with other areas of industry, pay is not necessarily a predictor of performance, however. Just as in the SOC, CTAC analysts must have ample opportunities for growth, autonomy, and recognition that their efforts make a difference.

To summarize our conversation about CTAC personnel needs, here are some hiring strategies for the CTAC:

1. Select from existing SOC staff—both when it is first formed and thereafter. Analysts who demonstrate desire and ability to learn malware analysis techniques, for instance, are good candidates for selection into a CTAC position.
2. When hiring from the outside, it is a good idea to integrate interview questions that emphasize problem solving, out-of-the-box thinking, and knowledge that crosses multiple disciplines. Warm-up questions like "What is the difference between TCP and UDP?" help weed out those who are clearly not qualified. On the other hand, breadth and depth of an answer to "What is your favorite [operating

system|network protocol|programming language] and why?" might reveal a lot about the candidate.

3. It is best not to overemphasize most certifications or knowledge of a specific tool. Knowing how to use a specific tool can usually be taught in a matter of days; knowing the theory behind how a tool functions, or which tool to use, takes much longer, and is usually harder to determine from a resume or a one-hour interview.

11.5.6 Physical Space

Estimating space requirements for a CTAC within a SOC is relatively straightforward. For however many people the CTAC will have once fully stood up, that's how many additional SOC cubicles and workstations are needed. The cubicles used by the CTAC are most like those used by Tier 2 analysts, with enough room to accommodate multiple monitors, the analysts' normal complement of workstations, potentially a malware analysis and/or forensics workstation, associated power and network cabling, a KVM, and a small collection of technical books. In addition, some members of the CTAC will need workspace for handling digital forensic artifacts, and hard copy of various cyber intel reporting. This space must also accommodate specialized forensics equipment, such as imagers, flyaway kits, and media safes.

Ideally, each section of the SOC (including the CTAC) will have its own clustered workspace. This may work to the SOC's advantage when standing up a CTAC; there may not be room near the SOC, meaning the CTAC will have to find a temporary room somewhere else in the same building as the SOC. Locating the CTAC more than a 5- or 10-minute walk from the SOC is not advised, because it will lead to isolation and discontinuity between people who must work together every day. Long term, the CTAC should be collocated with the rest of the SOC.

Assessing other non-personnel space needs for the CTAC is much more situation dependent. If the SOC is only able to keep a couple weeks of log data or PCAP data online, it will likely need additional physical rack space for servers and storage supporting the need for long-term historical query. Additional sensor equipment is also a major user of space, especially at remote locations. Finally, the SOC should plan for space to host an air-gapped lab capable of supporting malware analysis activities, if it doesn't already have one.

11.6 How Do We Get Started?

Creating a CTAC follows many of the same steps and timeframe as standing up a SOC (discussed in Appendix B), albeit in a somewhat compressed manner. Ideally, most of the groundwork for a CTAC already has been laid. External executive-level support for expansion of the SOC's roles and budget is typically necessary to stand up a CTAC. Common practices such as watching adversaries *in situ*, trading indicators of compromise, and detonating malware may be unsettling to some stakeholders. If enabling tools or logging aren't in place, substantial engineering support will probably be needed.

Any question that garnered a "No" from Section 11.4 should be dealt with in the first 12 to 18 months of the CTAC's existence. For this reason, SOCs that are less than 12 to 24 months old are advised against attempting to stand up a CTAC. The complexities of the SOC's tool suite, personnel, and governance structure are such that standing up a CTAC will likely stretch a young SOC's resources too thin, or depend upon capabilities that don't yet exist.

The following sections break down the different actions that will need to be accomplished in order to stand up a CTAC, assuming few or no existing CTAC capabilities are in place. While milestones and timelines can be adjusted, we offer an initial roadmap in terms of both priorities and sequencing.

11.6.1 Laying the Foundation: Month 0 and Prior

In this phase, our objectives are to establish the conditions under which a CTAC can achieve success. For some SOCs, this can take as little as a few months; for others it may take several years to lay the groundwork. Tasks include:

- Solicit support from upper management and executives.
- Secure funding for additions to SOC tools and personnel.
- Place initial space reservation for the CTAC.
- Ensure that most prerequisites in Section 11.4 have been met.
- Identify and draft any necessary governance and authorities to support major capabilities of the CTAC.
- Draft new organizational structure for the SOC reflecting the addition of the CTAC.

11.6.2 Build-out: 0 to 6 Months

In this phase, we begin constructing the CTAC, leveraging existing SOC personnel. Tasks include:

- Socialize, finalize, and receive approval for governance and authorities needed to support essential elements of CTAC operations.
- Identify existing resources (e.g., tools and analysts), that can be moved into the CTAC.
- Begin acquisition and engineering efforts for major CTAC tool needs.
- Perform bulk of hiring to support CTAC—either bringing analysts directly into the CTAC, or backfilling positions vacated by analysts who will move into the CTAC.
- Contract and execute facility build-out for CTAC workspace.

11.6.3 Initial Operations: 6 to 12 Months

In this phase, we can begin practicing the functions of CTAC. In its beginning, the CTAC will begin processing malware and cyber intel without much external impact. However, we should see the CTAC become its own entity with its own ops tempo and distinct identity. Tasks and milestones include:

- Formal establishment of the CTAC
- Authoring major CTAC SOPs
- Deployment and configuration of CTAC tool sets and analysis environment
- Beginning initial analysis duties and deliverables
- Beginning to collect open source intelligence and reaching out to other SOCs for indicator and intel sharing
- By the end of this phase, actively participating in threat-sharing groups and publishing some work product for consumption by trusted external parties
- Standup of a threat knowledge management tool or system, even if in an initial or prototype capacity.
- By the end of this phase, having at least 50 percent of the CTAC team onboard
- Starting to identify gaps in detection, response, and prevention capabilities, particularly additional or more detailed logging of key event types.

11.6.4 Sustained Capability: 12 to 18 Months

In the next phase of standup, our objective is to assume the full spectrum of capabilities initially planned for the CTAC. Tasks and milestones include:

- Finalize CTAC SOPs.
- At least 75 percent of CTAC members are onboard, and most other SOC positions vacated by moves to the CTAC are backfilled.
- Most core CTAC tools, such as forensics and malware analysis systems, are fully operational.
- Processing malware and cyber intel is a routine duty performed by multiple staff members.
- Aspects of malware analysis and cyber intel fusion ripe for automation are identified; analysts begin automating "low-hanging fruit."
- CTAC members are able to identify remaining gaps in foundational SOC tools that impact their mission, and are working with other SOC staff and engineers to plan how those gaps will be closed.
- A long-term solution for threat knowledge management is being built, if not already in place.

11.6.5 Long–Term Operations: 18+ Months

In the most mature phase of operations, we should to see the CTAC provide substantial return on investment. Milestones and major activities include:

- Regular threat briefings to cybersecurity stakeholders and executives become the norm.
- As TTPs and tools mature, the rate at which the CTAC is able to process malware samples and cyber intel, and scan for hits on threat indictors increases substantially.
- CTAC observations on adversary TTPs substantially influence incident response activities; "block/reformat/reinstall" is no longer an automatic response to all malware hits.
- The CTAC begins contemplating more in-depth adversary engagements, such as honeypotting, "fishbowl" architectures, and focused monitoring *in situ*.
- Resourcing and budgetary impacts related to CTAC operations and requirements stabilize to the point where they can be planned for in coming years.
- Every section of the SOC that performs monitoring or analysis duties uses the threat knowledge management tool during their daily routine.

There are a couple points worth noting. First, this timetable gives equal priority to cyber intel analysis and malware analysis functions; it is very likely that one capability will lead the other, depending on what capabilities existed prior to CTAC standup. Second, these timetables are notional. It may take some CTACs as few as 12 to 18 months to stand up, whereas others may still struggle after the three-year mark. Nothing changes overnight, and getting into a consistent operations tempo and rhythm generally takes *years* no matter how talented the SOC's analysts are.

11.7 Peering

One of the best things a SOC can do to get ahead in the world is to make friends with other SOCs. These relationships are typically initiated through mutual membership in industry or professional associations, geographic proximity, voluntary membership in a SOC association such as Forum for Incident Response and Security Teams (FIRST) [275] or various Information Sharing and Analysis Centers (ISACs) [276], conference attendance, or through a national SOC organization. Collaboration is not something that can be mandated or ordered by a coordinating SOC; rather it is built upon direct analyst-to-analyst communication, mutual respect, and a quid-pro-quo attitude—all of which may take months and years to flourish.

SOCs have a lot to offer one another, including:

- "Heads up" on recently observed anomalous and malicious activity

- Useful defender TTPs and CONOPS
- Cyber observables, indicators of compromise, adversary TTPs, and incident tips
- IDS and SIEM signatures, content, and use cases
- Custom tools and scripts.

Consider, for instance, a situation where a SOC compiles a large amount of cyber intel into a list of IDS signatures or SIEM content. This process may take months or years of analyst work and is an ongoing effort. By pooling these signatures and sharing processed cyber intel with other SOCs, many participants stand to save a tremendous amount of time, while simultaneously instrumenting their systems against a much wider set of threats.

More important, collaboration between SOCs builds a sense of partnership and belonging that the SOC is unlikely to get from its own constituency. As we discussed in Section 5.2, a SOC is likely to feel isolated. Through collaboration with other SOCs, it is likely to find validation in its challenges and hints for future success. Also, through collaboration and sharing, a SOC is better able to gauge its competency and maturity, especially when it is subject to external audit or scrutiny. If there is anything a SOC can do to get ahead for free, it very well may be contact and frequent collaboration with its peers.

For more information on inter-SOC information sharing, see Chapter 6 of [15] and [277].

11.8 Where to Get Cyber Intel and What to Do with It

World-class SOCs have at their disposal a virtual river of cyber intel. Not all cyber intel is created equal, and SOCs must be careful about what cyber intel to favor in driving updates to monitoring and analytics. Good cyber intel is:

- **Timely**, often describing events that happened only minutes, hours, or days ago
- **Relevant** to the recipients' environment and threats
- **Accurate**, which is to say its content correctly describes what happened
- **Specific** about the incident, observable, or adversary it describes
- **Actionable**, such that the recipient can do something constructive with it.

Some of the best cyber intel a SOC can get is from other peer SOCs with which it has a direct, personal, trusted relationship. In such cases, the recipients are well aware of the quality of the material they are getting and are likely to get it very quickly, and the intel is likely subject to minimal redaction as there was no intermediary to "water down" its content.

That said, direct relationships are built up over time, and fledgling SOCs often must first go to open sources to get started, such as various organizations that publish material on the Internet. These sources are large in number, voluminous in content, and overlap very frequently. However, there is no "standard" list of cyber news and cyber intelligence sources. Therefore, to get readers started, here is a list of links that a SOC may find useful to visit on a regular basis:

Internet health:

- ISC: http://www.isc.org
- NetCraft: http://news.netcraft.com/
- US-CERT: http://www.US-Cert.gov

General technology and security trends:

- Schneier on Security Blog: http://www.schneier.com/
- Krebs on Security: http://krebsonsecurity.com/
- Security Dark Reading: http://www.darkreading.com/
- Slashdot: http://slashdot.org
- Engadget: http://www.engadget.net
- Securosis: https://securosis.com/blog

Threat intelligence:

- Microsoft Security Intelligence Report: http://www.microsoft.com/security/sir/default.aspx
- Team Cymru (also has subscription service): www.team-cymru.org
- FBI Cybercrime information: http://www.fbi.gov/about-us/investigate/cyber/cyber

Malware and threats:

- Threat Expert: http://threatexpert.com
- Microsoft Malware Protection Center: http://www.microsoft.com/security/portal/default.aspx
- SANS Internet Storm Center: http://Isc.sans.edu
- Symantec Threat Explorer: http://www.symantec.com/norton/security_response/threatexplorer/index.jsp
- Symantec Internet Threat Report: http://www.symantec.com/business/theme.jsp?themeid=threatreport
- McAfee Threat Center: http://www.mcafee.com/us/threat_center/
- Metasploit Blog: https://community.rapid7.com/community/metasploit?view=blog
- Security Focus: http://www.securityfocus.com/
- Dshield: http://www.dshield.org/
- Offensive Security's Exploit Database: http://www.exploit-db.com/
- Worldwide Observatory of Malicious Behaviors and Attack Threats (WOMBAT): http://wombat-project.eu/

- Symantec's Worldwide Intelligence Network Environment (WINE): http://www. symantec.com/about/profile/universityresearch/sharing.jsp
- Mandiant M-Trends: https://www.mandiant.com/resources/mandiant-reports/

Bad domains, IP addresses, and other indicators:
- Malware Domain Blocklist: http://www.malwaredomains.com/
- Malware Domain List: http://www.malwaredomainlist.com/
- Unspam Technologies Project Honeypot: http://www.projecthoneypot.org/index.php
- EXPOSURE (Exposing Malicious Domains): http://exposure.iseclab.org/
- Shadowserver Foundation: http://www.shadowserver.org/wiki/
- Any of the other sources listed on page 13 of [235]

Automatic threat analyzers:
- Anubis (Analyzing Unknown Binaries): http://anubis.iseclab.org/
- Virustotal: http://www.virustotal.com/
- Metascan online: http://www.metascan-online.com/

Threats with signatures:
- IBM ISS X-Force: http://xforce.iss.net
- BotHunter Internet Distribution Page: http://www.bothunter.net/
- Latest Snort publicly available Snort rules (most recent rules require subscription): http://www.snort.org/snort-rules/
- Emerging Threats signature list: http://www.emergingthreats.net/
- Latest Tenable Nessus plugins (requires subscription): http://www.nessus.org/plugins/

Patches and vulnerabilities:
- MITRE's CVE: http://cve.mitre.org
- NIST's National Vulnerability Database: http://nvd.nist.gov/
- US-CERT Technical Cyber Security Alerts: http://www.us-cert.gov/cas/techalerts
- Microsoft Security TechCenter: http://technet.microsoft.com/en-us/security/default.aspx
- Whatever other vendor software is commonly used within the constituency.

In addition to these open sources, some SOCs leverage commercial feeds of cyber intel such as Bit9 [278], CrowdStrike [279], iDefense [280], and Symantec DeepSight [281]. Finally, SOCs can share information on a pairwise or group basis, leveraging STIX (Structured Threat Information Sharing) framework, TAXII (Trusted Automated eXchange of Indicator Information) protocols and services, and CybOX (Cyber Observables eXpression) language [282].

On the basis of the variety of cyber intel and news it consumes and its capabilities, the SOC has a number of options for synthesis and redistribution both internally and externally:

- Critical patch notices to the constituency
- Daily, weekly, or monthly cybersecurity newsletters or digests
- Signature tuning and signature updates
- Custom IDS signatures
- Custom SIEM content
- Ad hoc, targeted, or short-term analysis taskings to Tier 1 or Tier 2 analysts
- Redistribution, perhaps in a common machine-readable format, to other SOCs
- Tactical or strategic threat assessments of adversary TTPs.

Most established SOCs will publish some kind of regular cybersecurity newsletter or digest to cybersecurity stakeholders in the constituency. While this function may obligate a quarter- or half-staff FTE, it gives the SOC "good press" and ensures the SOC is keeping up with the news. More important, however, it demonstrates to the constituency that the SOC is the go-to shop for cyber SA.

When looking at any kind of custom IDS signatures or SIEM content, the SOC has a number of indicators to scrape from bad domain lists and incident reporting. These may include:

- IP addresses
- Domain names
- Network traffic content
- Email addresses
- User names
- File names
- File hashes.

When these indicators trigger a signature hit on an IDS alert or audit data feed (such as Web proxy or firewall logs), it could indicate a couple of things, depending on which stage of the cyber attack life cycle is of interest. The bad IP addresses and domain names are often used to indicate RATs, malware beaconing, botnet command and control, or data exfiltration. Triggering on a file name or file hash, especially on an email attachment, might indicate a phishing attack from a known adversary or some other known piece of malware. The point is that the response actions would be very different, depending on the type of indicator that was matched. So the first thing that SOC analysts must understand before feeding indicators into their data analytics is what to do when they fire.

The SOC should, however, be cautious when scraping indicators. Cyber intel authors sometimes put disclaimers in their reports, saying something to the effect of "do not block activity based on these indicators." They may or may not articulate their reasons for doing so, but it is important for the receiving SOC not to break the trust of the originating SOC in using those indicators to set up blocks. It is often better to use collected cyber intel for passive monitoring. Second, scraping entire lists such as malware domain block lists is tempting but can be error prone for the following reasons:

- The domains are likely to be very short-lived, such as the case with fast-flux networks [283] and, therefore, might be useless by the time they are detected. The SOC must remain vigilant to keep its bad indicator lists up-to-date, carefully aging off indicators that are no longer likely to be seen. This can have benefits from a performance perspective, as well as minimizing false positives.

- In a list of 100,000 or more indicators, there may not be any differentiation between high- and low-priority items. If a sensor hits on any one of them, is that cause for serious concern? The SOC may want to be choosy in what sources it uses.

- Some indicators such as file names are very prone to false positives due to the lack of uniqueness of the name (e.g., don't alert every time report.doc is seen in an email attachment). On the other hand, others may never alert (e.g., file hashes) because they change very often and, in some cases, are computationally expensive to use.

- Because of the trend toward third-party hosting services and cloud computing, it is sometimes very ambiguous whether a given IP address or domain is "bad" or not. A given Web-hosting company may have a dozen "bad" domains mixed in a subnet with a thousand other websites. Alerting (or blocking) the entire subnet just for a few bad domains doesn't always make sense. Also, a given website may be hosted in five different places. Does an indicator match against only one hosting location or all five?

- When looking at human-readable, narrative-style cyber intelligence reports and incident reporting, there may be victim indicators such as IP addresses mentioned, as well as the attacker indicators. This makes automated harvesting of indicators problematic because the good indicators could easily be swept up with the bad (leading to low-quality indicator databases and false positives down the line).

- There is often tremendous overlap between many indicator lists, depending on their source. SOCs leveraging multiple sources should be careful to deduplicate indicator lists before using them to instrument monitoring systems.

In short, leveraging indicator lists when instrumenting automated analytic engines (such as the SIEM) demonstrates many of the same problems and pitfalls as anything else that is signature based. We must be very careful to maximize our true positive rate while minimizing false positives. As a result, some SOCs choose to harvest indicators only from detailed security incident reports authored by other SOCs with which they have a trusted relationship. This allows the SOC to make informed decisions about the severity and criticality of the associated intrusion activity at the time it harvests the indicators. However, this approach is highly dependent upon having a large pool of recent reports from partner SOCs, and the analyst cycles to process them.

Chapter 12

STRATEGY 10: Stop. Think. Respond . . . Calmly

....

In a given year, SOCs will track hundreds or even thousands of cases, vulnerabilities, and threats. In each instance, the SOC must render a response that is appropriate, given the criticality of the situation. As a result, the majority of our incident handling should be routine and not cause for an emergency. In our tenth and last strategy, we examine techniques for addressing incidents in a professional, trustworthy, and effective manner. Accordingly, we discuss how to track incidents from cradle to grave.

12.1 Professional Incident Response

When there is a major incident, all eyes are on the SOC. If it has followed the guidance laid out in the previous sections of this book, most aspects of incident handling should come naturally. The SOC should have the following in place:

- A workforce with strong technical, analytic, and communication skills
- CONOPS, SOPs, and escalation procedures that guide the SOC's actions
- Means to coordinate analysis and response activity among members of the SOC
- Established POCs with whom to coordinate response actions
- Established and ad hoc log, PCAP, and live system image data collection and analysis tools sufficient to help establish the facts about incidents

- The authorities to enact swift and decisive response actions when called for and passive observation or incident de-escalation when they are not.

We must ensure our incident response is efficient, effective, relevant, and complete. Failure to do so could undermine the SOC mission, which is to limit damage, assess impact, and render a durable response. Let's consider some dos and don'ts when we think the SOC has found something bad:

- **Follow your SOPs.**

  No two incidents are exactly the same, and some are more complex than others. That said, most incident handling should be routine—easily handled by one or two analysts—and no great cause for concern. They should fall under well-structured SOPs that can be picked up by members of the SOC and easily understood. This saves the SOC's energy for cases that fall outside the daily routine, such as root compromises, whose response is not entirely formulaic and cannot be completely scripted.

- **Don't panic.**

  When police, firefighters, or paramedics arrive on the scene of a 911 call, they are cool, calm, and collected. They are able to assess and stabilize the situation and direct response accordingly. Doing so engenders trust on the part of the complainant or the victim. The SOC should follow the same practice. For those not familiar with CND operations, an incident is cause for great excitement and emotion. This can lead to reactions that amplify damage. The SOC will gain the trust of those involved if it provides measured response, no matter what circumstances it encounters.

- **Don't jump to conclusions.**

  "Oh my god, we're being attacked!" has been uttered in response to many an incident. Are we really? What is causing us to draw this conclusion? Are we just looking at IDS alerts, or do we have a system image that clearly indicates a root compromise? It takes a skilled analyst to correctly interpret what a set of security logs or media artifacts do or do not say. Recognizing the limits of our understanding of a situation is critical, especially when an unambiguous "smoking gun" is hard to find.

- **Be careful about attribution.**

  A NetFlow record may indicate that an entity from Kazblockistan is scanning our enterprise or is receiving DNS beaconing from a compromised host. Is it really someone in that country or is that just the next hop out in the network connection? Furthermore, just because an audit log is stamped with user Alice, was it really Alice sitting at the keyboard, was it Trudy who compromised Alice's account, or, perhaps was it automated activity using Alice's identity? Most times, an incident responder can only propose theories and suggest a degree of confidence about who is behind a given set of malicious or anomalous activities. Unless we can actually prove who is sitting at the keyboard, user attribution is theory and not fact.

- **Assess the full extent of the intrusion.**
  We have a malware hit against a box—Was it the only one compromised? We see a privilege escalation attack on a given system—Is this box linked through a trust relationship to other systems or enclaves? We found some malware on a box involved in a compromise—What other indicators can we find that point to what activity, by whom, and at what stage in the attack life cycle? Shallow analysis can be very dangerous, and the operator must endeavor to understand the full scope of what has occurred. *Gather as much relevant evidence as possible and exploit it to the maximum extent practicable.* This goal must, of course, be balanced with the need to act in a timely manner, even though you don't have *all* of the facts nailed down.

- **Understand the "so what?"**
  When the SOC explains an incident to stakeholders and upper management, the bottom line is not about bits and bytes, it's about mission, dollars, and, sometimes, lives. The SOC must translate technical jargon into business language. There are four questions that should be answered: (1) what (and/or who) was targeted, (2) was the adversary successful, (3) who is the adversary and what is its motivation, and (4) how do we continue the mission?

- **Follow rules of evidence collection and documentation, when appropriate.**
  The more critical the incident, the greater pressure the SOC will likely face. All too often, the SOC must draw both a timeline of the adversary's actions and a timeline of how the SOC responded. By carefully documenting its incidents and incident handling, the SOC can demonstrate the rigor behind its actions, when scrutinized. Documenting everything also means clearly having incident evidence in careful order. Finally, in the case of collecting artifacts and documenting actions taken, the SOC must carefully follow any applicable digital forensics or evidence collection laws for their jurisdiction. In fact, it often is best to err on the side of having forensically sound evidence, even when the SOC doesn't initially think the case has any legal significance.

- **Provide measured updates at measured times.**
  When a hospital patient goes in for surgery, family members sit for hours in a waiting room, anxiously awaiting news of their loved one's fate. While it would be great to hear frequent updates on their loved one's procedure, doing so would impede the surgeon's ability to complete the operation correctly and in a timely manner. When firefighters show up at the scene of a fire, the onsite incident commander calls the

shots. The district fire chief and the city mayor generally don't show up because there is no need. For most SOCs, these clear boundaries of trust and communication are not as well established as for doctors or emergency responders.

In cyber incident response, the SOC must play a careful balancing act between keeping management and constituents up-to-date and executing analysis and response efforts. If not careful, key analysts will constantly be pulled away from actually analyzing and responding in order to brief stakeholders. It is wise for SOC leadership to manage expectations of constituency seniors and run interference so the SOC can continue with the mission.

During a serious incident, the SOC may consider two separate regular meetings every day or two. The first is for direct players in the incident who can talk bits and bytes, and usually occurs informally on the SOC ops floor or over the phone. The second is a more formal SA update to upper management. This keeps seniors out of the weeds, ensures everyone is on the same page, and allows SOC personnel to focus on operations.

The SOC should also be careful about which parties are given status updates. Many parties want to know about every incident that leaves the SOC, yet, in many cases, their need to know is tenuous at best. The SOC can cut down on second-guessing and time spent reporting status to external parties by carefully negotiating a reporting structure for major incident types.

In addition, it's important to let junior members of the SOC team know that they are not to release details on the incident without authorization. A SOC's credibility can be easily destroyed by just one or two cases where a Tier 1 analyst picked up the phone and gave "half-baked" incident details to the wrong constituent. Furthermore, the SOC must be careful not to let details of incidents leak out in emails or other communications that could be seen by an adversary.

- **Carefully assess the impact of countermeasures and response actions.**
  The SOC must work with system owners and sysadmins in order to get to the bottom of an incident through careful artifact collection, analysis, and damage assessment. The SOC should not perform "knee-jerk" response actions that may take down key mission systems or networks. Blindly reimaging and reinstating systems involved in an incident without performing artifact and malware analysis is almost always counterproductive, because (1) we don't know whether the adversary has lost its foothold, and (2) we will never be able to fully assess what actually happened.

  Rather, the SOC must understand how proposed countermeasures will impact their ability to assess the extent of the intrusion and how the adversary's actions might change as a result. SOCs that have strong adversary engagement skills may actually enact a series of response measures designed to guide the adversary toward a desired goal, revealing additional details of the adversary's TTPs and motives.

- **Ensure the entire SOC is working toward the same goal.**
  In the heat of the moment, it is easy for members of the SOC to step beyond what they are authorized to do, considering their limited perspective on what needs to happen next. Telling a system owner to disconnect a system or shut off access could be disastrous, even if it seems like the right thing to do at the time. Coordination isn't just between the SOC and external parties—it starts internally, through both peer-to-peer collaboration and a clear command structure.

- **Don't be afraid to ask for help.**
  Not every SOC has all the skills and knowledge in-house to handle every intrusion. Incidents must be evaluated within the context of the mission and systems they impact—meaning the SOC must frequently reach out to system owners. Is an attack targeted at a specific business or geographic region? By talking to other partners, the SOC can find out more. Do we have the necessary skills to analyze a piece of malware? If not, another SOC or third party might provide reverse engineering expertise.

12.2 Incident Tracking

Every mature SOC needs a robust incident tracking capability. However, there is no one size fits all, meaning every SOC does it a little differently. In this section, we talk about key requirements and architectural options for incident tracking. We also discuss areas in which an incident tracking system (by itself) falls short, indicating the SOC should seek out additional forms of knowledge management.

The SOC's needs for incident tracking are not terribly different from general case ticketing and tracking used in general IT help desk and system administration. That said, the SOC has several key requirements, many of which are unique to CND:

- Allows consistent and complete information capture across incidents for each state of the incident life cycle—Tier 1 triage, Tier 2 analysis, response, closure, and reporting
- Is able to record both structured information from analysts (incident category, time reported), semi-structured data (impacted users, impacted systems) and non structured information (analyst narrative), along with time-stamped notes
- Is available to SOC personnel while protecting sensitive details from constituents, thereby avoiding compromise of any insider threat cases or word getting out about an incident prematurely or to the wrong parties
- Protects details about cases even if the general constituency is compromised

- Supports escalation and role-based access control for different sections within the SOC
- Supports long-term trending and metrics
- Can incorporate artifacts or pointers to artifacts, such as events or malware samples.

It's also important to note that as an alternative to calling the SOC or sending it an email, constituents could input information about suspected cyber incidents on a form on the SOC website. Although this information might automatically populate a case, submitters outside of a SOC should not have access to that information after it is submitted (unlike an IT trouble ticketing system). Many SOCs will choose to keep this form submission system separate from their internal case system for security purposes.

Unfortunately, there is no standard IT case management system used for CND. Usually, SOCs will adopt one approach from those listed in the Table 24. Let's look at the pros and cons of each potential approach to cybersecurity incident tracking.

Table 24. Case Tracking Approaches

Approach	Pros	Cons
Manually (on paper). Each case is tracked through a collection of hard-copy notes and artifacts.	Free Easy to set up and use Escalation is straightforward. Compromise of SOC systems does not compromise case data.	Can be slow. Paper copies can be lost. Large amounts of paper accumulate over time. Lack of structured forms can lead to inconsistent tracking, especially over time. Very "19th century" Trending or metrics are manual.
Manually (in soft copy). The same as hardcopy, but artifacts are left on a network share.	Startup is relatively straightforward, assuming SOC already has a file share.	Nearly as haphazard as hardcopy Lack of structured forms can lead to inconsistencies over time. Trending or metrics are manual. Short of changing directory and file permissions to each case, loss or compromise of data is possible.
Existing IT case management system	Acquisition and O&M free to the SOC. Reporting and metrics possible. Seamless escalation of cases from/to IT help desk	Unlikely to be flexible to SOC needs. Sensitive data is comingled with general IT help tickets. Ticketing sysadmins, power users can see SOC's cases; very high likelihood of compromise of internal threat cases. If general constituency systems are compromised, it is fair to assume the adversary can see SOC cases. Incorporating case artifacts may be a challenge.

Table 24. Case Tracking Approaches

Approach	Pros	Cons
SOC instance of COTS IT case management system	System comes with polished feature set, documented setup, and central administration. Robust reporting and metrics possible Case details available only to parties designated by SOC	Usually very expensive Customization to SOC needs might be a challenge. Incorporating case artifacts may be a challenge.
SOC instance of FOSS IT case management system	Depending on tool chosen, system comes with polished feature set, documented setup, and central administration. Very flexible Free to acquire Reporting and metrics possible Case details available only to parties designated by SOC	General IT case tracking system will require nontrivial customization to fit CND use cases; not "turnkey." O&M customization will likely require staff with some experience in scripting, programming, or databases.
Custom–built ticketing system	Extremely flexible Reporting and metrics possible Case details available only to parties designated by SOC	Expensive to acquire and O&M SOC must build system from scratch, requiring staff with extensive experience with programming and databases. Development of system may take a while, since SOC must start from scratch.
SIEM case tracking system	Free if SOC owns a SIEM System is specifically built for tracking security incidents. System leverages user groups and permissions setup for other SIEM tasks. Users can attach events and some artifacts to tickets. Escalation paths are useful if SOC leverages an event–driven workflow and correlation rules. Reporting and metrics possible	Extremely expensive if SOC does not own a SIEM Limited to no flexibility, depending on SIEM product If SIEM goes down, nearly all aspects of SOC operations (triage, analysis, case tracking) are kaput.

There are a number of issues to consider here. Some SOCs will get started with a manual hardcopy or softcopy case management system, but as we can imagine, this will not last for very long. Leveraging an existing IT help desk ticketing system may also seem appealing, but the SOC has a specific set of needs and sensitivities. As such, that option isn't very appealing either. A few SOCs have been known to build their own custom ticketing system from scratch, but only when requirements and customized use cases support the resulting expense, as is sometimes the case in large, tiered, and coordinating SOCs. There may be good examples where case-by-case access controls are needed, such as with a SOC that has as strong law enforcement or insider threat mission need.

On the basis of the pros and cons, many mature SOCs choose to leverage their own customized instance of a FOSS ticketing system. Request Tracker (RT) has been openly customized for use by SOCs [284], making it an appealing option. If the SOC chooses to implement or customize its own ticketing system, there are a number of existing examples of incident tracking forms found in Section 3.7.5 of [3].

That said, best-of-breed SIEMs have complex correlation capabilities. With these they can automatically generate cases prepopulated with key information. This is commonly used for cut-and-dried incidents like AV hits. Implementing automatic case generation for these use cases can save Tier 1 and Tier 2 a lot of time, but may be contingent on using the SIEM's ticketing system. One alternative is to have the SIEM automatically send event details in a scripted action to a customized FOSS ticketing system. The critical decision here is where the SOC chooses to bring the analysts' workflow out of the SIEM and into a third-party ticketing system.

As mentioned in Section 11, it's also important to recognize that not everything a SOC needs to track over time may be contained in case notes. For instance, the SOC will likely want to build a knowledge base that is system-, adversary-, or TTP-focused, rather than case-focused. Some SIEMs have internal knowledge base features, but such functionality tends to be limited in its customizability. For more information on cyber threat knowledge management, see Section 11.

References

[1]   Wikimedia Foundation, Inc., "Advanced Persistent Threat," 3 Feb 2014. [Online]. Available: http://en.wikipedia.org/wiki/Advanced_persistent_threat. [Accessed 13 Feb 2014].

[2]   R. G. Bace, Intrusion Detection, Indianapolis: Macmillan Technical Publishing, 2000.

[3]   G. Killcrece, K.-P. Kossakowski, R. Ruefle and M. Zajicek, "State of the Practice of Computer Security Incident Response Teams (CSIRTs)," October 2003. [Online]. Available: http://resources.sei.cmu.edu/library/asset-view.cfm?assetID=6571. [Accessed 13 Feb 2014].

[4]   Killcrece, Georgia; Kossakowski, Klaus-Peter; Ruegle, Robin; Zajicek, Mark, "Organizational Models for Computer Security Incident Response Teams," December 2003. [Online]. Available: www.cert.org/archive/pdf/03hb001.pdf. [Accessed 13 Feb 2014].

[5]   S. Northcutt, Network Intrusion Detection (3rd Edition), Indianapolis: New Riders Publishing, 2002.

[6]   T. Parker, E. Shaw, E. Stroz, M. G. Devost and M. H. Sachs, Cyber Adversary Characterization: Auditing the Hacker Mind, Rockland, MA: Syngress Publishing, Inc., 2004.

[7]   L. Spitzner, Honeypots: Tracking Hackers, Addison-Wesley Professional, 2002.

[8]   M. J. West-Brown, D. Stikvoort, K.-P. Kossakowski, G. Killcrece, R. Ruefle and M. Zajicekm, "Handbook for Computer Security Incident Response Teams (CSIRTs)," April 2003. [Online]. Available: http://resources.sei.cmu.edu/library/asset-view.cfm?assetid=6305. [Accessed 13 Feb 2014].

[9]   R. Bejtlich, The Tao of Network Security Monitoring: Beyond Intrusion Detection, Boston, MA: Pearson Education, 2005.

[10]   K. R. Van Wyk and R. Forno, Incident Response, Sebastopol, CA: O'Reilly Media, Inc., 2001.

[11]   C. Prosise, K. Mandia and M. Pepe, Incident Response and Computer Forensics, Second Edition, McGraw-Hill/Osborne, 2003.

[12]   E. E. Schultz and R. Shumway, Incident Response: A Strategic Guide to Handling System and Network Security Breaches, Sams, 2001.

[13]   R. Bejtlich, Extrusion Detection: Security Monitoring for Internal Intrusions, Addison-Wesley Professional, 2005.

[14]   C. Fry and M. Nystrom, Security Monitoring, Cambridge: O'Reilly, 2009.

[15]   D. Rajnovic, Computer Incident Response and Product Security, Indianapolis, IN: Cisco Press, 2011.

[16]   C. Sanders and J. Smith, Applied Network Security Monitoring: Collection, Detection, and Analysis, Boston, MA: Syngress, 2013.

[17]   J. Carr, Inside Cyber Warfare, Sebastopol, CA: O'Reilly Media, 2010.

[18]   J. Andress and S. Winterfeld, Cyber Warfare, Waltham, MA: Elsevier, Inc., 2011.

[19]   P. Cichonski, K. Masone, T. Grance and K. Scarfone, "NIST Special Publication 800-61 Rev 2: Computer Security Incident Handling Guide," August 2012. [Online]. Available: http://csrc.nist.gov/publications/nistpubs/800-61rev2/SP800-61rev2.pdf. [Accessed 13 Feb 2014].

[20]   R. Bejtlich, Practice of Network Security Monitoring, San Francisco, CA: No Starch Press, 2013.

[21]   W. R. Stevens and K. R. Fall, TCP/IP Illustrated, Vol. 1: The Protocols (2nd Ed.), Boston: Addison-Wesley Professional, 2011.

[22]   G. R. Wright and W. R. Stevens, TCP/IP Illustrated, Vol. 2: The Implementation, Boston: Addison-Wesley Professional, 1995.

[23]   C. Sanders, Practical Packet Analysis: Using Wireshark to Solve Real-World Network Problems, San Francisco: No Starch Press, 2011.

[24]   C. Steel, Windows Forensics: The Field Guide for Corporate Computer Investigations, Hoboken: Wiley, 2006.

[25]   H. Carvey, Windows Forensic Analysis Toolkit, Third Edition: Advanced Analysis Techniques for Windows 7, Waltham: Syngress, 2012.

[26]   K. J. Jones, R. Bejtlich and C. W. Rose, Real Digital Forensics: Computer Security and Incident Response, Boston: Addison-Wesley Professional, 2005.

[27]   B. Carrier, File System Forensic Analysis, Boston: Addison-Wesley Professional, 2005.

[28]   D. Farmer and W. Venema, Forensic Discovery, Boston: Addison-Wesley Professional, 2005.

[29]   C. L. Brown, Computer Evidence: Collection and Preservation, Boston: Course Technology, 2009.

[30]   P. Engebretson, The Basics of Hacking and Penetration Testing: Ethical Hacking and Penetration Testing Made Easy, Syngress, 2011.

[31]   D. Stuttard and M. Pinto, The Web Application Hacker's Handbook: Finding and Exploiting Security Flaws, Hoboken, NJ: Wiley, 2011.

[32]   J. Scrambray, V. Liu and C. Sima, Hacking Exposed Web Applications, 3rd Ed, New York, NY: McGraw-Hill Osborne Media, 2010.

[33]   C. Anley, J. Heasman, F. Lindner and G. Richarte, The Shellcoder's Handbook: Discovering and Exploiting Security Holes, Hoboken: Wiley, 2007.

[34]   D. Litchfield, C. Anley, J. Heasman and B. Grindlay, The Database Hacker's Handbook: Defending Database Servers, Hoboken: Wiley, 2005.

[35]   P. Herzog, "Open Source Security Testing Methodology Manual (OSSTMM)," 2014. [Online]. Available: http://www.isecom.org/research/osstmm.html. [Accessed 13 Feb 2014].

[36]   M. Sikorski and A. Honig, Practical Malware Analysis, San Francisco: No Starch Press, 2012.

[37] M. Ligh, S. Adair, B. Hartstein and M. Richard, Malware Analyst's Cookbook and DVD: Tools and Techniques for Fighting Malicious Code, Hoboken: Wiley, 2010.

[38] C. Eagle, The IDA Pro Book: The Unofficial Guide to the World's Most Popular Disassembler, San Francisco: No Starch Press, 2011.

[39] C. H. Malin, E. Casey and J. M. Aquilina, Malware Forensics: Investigating and Analyzing Malicious Code, Waltham: Syngress, 2008.

[40] E. Eilam, Reversing: Secrets of Reverse Engineering, Hoboken: Wiley, 2005.

[41] NSS Labs, Inc., "Test Reports | NSS Labs," 15 Jan 2014. [Online]. Available: https://www.nsslabs.com/reports/categories/test-reports. [Accessed 13 Feb 2014].

[42] Committee on National Security Systems, "CNSS Instruction No. 4009," Committee on National Security Systems, Ft. Meade, 2010.

[43] N. Brownlee and E. Guttman, "Request for Comments 2350 Expectations for Computer Security Incident Response," June 1998. [Online]. Available: http://www.ietf.org/rfc/rfc2350.txt. [Accessed 20 Feb 2012].

[44] R. Shirey, "RFC4949, Internet Security Glossary, Version 2," August 2007. [Online]. Available: http://tools.ietf.org/html/rfc4949. [Accessed 13 Feb 2014].

[45] NIST, "Information Security Continuous Monitoring (ISCM) for Federal Information Systems and Organizations, NIST SP 800-137," September 2011. [Online]. Available: http://csrc.nist.gov/publications/nistpubs/800-137/SP800-137-Final.pdf. [Accessed 13 Feb 2014].

[46] The SANS Institute, "Security Training Courses," 2012. [Online]. Available: http://www.sans.org/security-training/courses.php. [Accessed 5 Apr 2012].

[47] M. R. Endsley, "Toward a Theory of Situation Awareness in Dynamic Systems," *Human Factors,* pp. 32–64, 1 March 1995.

[48] R. Marty, Applied Security Visualization, Boston, MA: Pearson Education, Inc., 2009.

[49] G. Conti, Security Data Visualization: Graphical Techniques for Network Analysis, San Francisco, CA: No Starch Press, 2007.

[50] D. A. D'Amico, "VizSec.org," 30 Jan 2014. [Online]. Available: http://www.vizsec.org/. [Accessed 13 Feb 2014].

[51] J. Boyd, "The OODA Loop," Feb 2010. [Online]. Available: http://www.danford.net/boyd/essence4.htm. [Accessed 13 Feb 2014].

[52] E. M. Hutchins, M. J. Cloppert and R. M. Amin, "Intelligence-Driven Computer Network Defense Informed by Analysis of Adversary Campaigns and Intrusion Kill Chains," Proceedings of the 6th International Conference on Information-Warfare & Security (ICIW 2011), March 2011. [Online]. Available: http://www.lockheedmartin.com/content/dam/lockheed/data/corporate/documents/LM-White-Paper-Intel-Driven-Defense.pdf. [Accessed 7 October 2012].

[53] National Security Agency, "Defense in Depth," 8 June 2001. [Online]. Available: http://www.nsa.gov/ia/_files/support/defenseindepth.pdf. [Accessed 13 Feb 2014].

[54]   D. J. Castello, "There's Nothing Wrong With Our Radar!," 2001. [Online]. Available: http://www.pearl-harbor.com/georgeelliott/index.html. [Accessed 13 Feb 2014].

[55]   S. Axelsson, "The Base-Rate Fallacy and its Implications for the Difficulty of Intrusion Detection," in *Recent Advances in Intrusion Detection*, 1999.

[56]   National Oceanic and Atmospheric Administration, "NOAA Deepwater Horizon Archive," 2014. [Online]. Available: http://www.noaa.gov/deepwaterhorizon/. [Accessed 13 Feb 2014].

[57]   D. S. Hilzenrath, "Technician: Deepwater Horizon Warning System Disabled," *The Washington Post*, 23 July 2010.

[58]   Cable News Network, "Fortune 500," 2014. [Online]. Available: http://money.cnn.com/magazines/fortune/fortune500/. [Accessed 13 Feb 2014].

[59]   Forbes.com LLC, "The World's Biggest Public Companies," 2014. [Online]. Available: http://www.forbes.com/global2000/list/. [Accessed 13 Feb 2014].

[60]   G. Killcrece, "Steps for Creating National CSIRTs," August 2004. [Online]. Available: http://resources.sei.cmu.edu/library/asset-view.cfm?assetid=53062. [Accessed 13 Feb 2014].

[61]   The SANS Institute, "Information Security Policy Templates," 2014. [Online]. Available: http://www.sans.org/security-resources/policies/. [Accessed 13 Feb 2014].

[62]   Wikimedia Foundation, Inc., "T-shaped skills," 27 Nov 2013. [Online]. Available: http://en.wikipedia.org/wiki/T-shaped_skills. [Accessed 13 Feb 2014].

[63]   Sourcefire, Inc., "Snort :: Home Page," 2014. [Online]. Available: http://www.snort.org/. [Accessed 13 Feb 2014].

[64]   Open Information Security Foundation, "Suricata," 2014. [Online]. Available: http://suricata-ids.org/. [Accessed 13 Feb 2014].

[65]   The Bro Project, "The Bro Network Security Monitor," 2014. [Online]. Available: http://www.bro.org/. [Accessed 13 Feb 2014].

[66]   QoSient, LLC, "Argus," 20 Jan 2014. [Online]. Available: http://www.qosient.com/argus/. [Accessed 13 Feb 2014].

[67]   Carnegie Mellon University, "CERT netSA Security Suite," 2014. [Online]. Available: http://tools.netsa.cert.org/. [Accessed 13 Feb 2014].

[68]   D. Miessler, "A tcpdump Tutorial and Primer," 2014. [Online]. Available: http://danielmiessler.com/study/tcpdump/. [Accessed 13 Feb 2014].

[69]   Wireshark Foundation, "Wireshark. Go Deep.," 2014. [Online]. Available: http://www.wireshark.org/. [Accessed 13 Feb 2014].

[70]   McAfee, Inc., "McAfee Network Security Platform," 2014. [Online]. Available: http://www.mcafee.com/us/products/network-security-platform.aspx. [Accessed 13 Feb 2014].

[71]   McAfee Inc., "McAfee ePolicy Orchestrator," 2014. [Online]. Available: http://www.mcafee.com/us/products/epolicy-orchestrator.aspx. [Accessed 13 Feb 2014].

[72] Hewlett-Packard Development Company, L.P., "Network Security | HP Official Site," 2014. [Online]. Available: http://www8.hp.com/us/en/software-solutions/network-security/index.html. [Accessed 13 Feb 2014].

[73] Hewlett-Packard Development Company, L.P., "SIEM, Information Security | HP Official Site," 2014. [Online]. Available: http://www8.hp.com/us/en/software-solutions/siem-arcsight/. [Accessed 13 Feb 2014].

[74] Splunk, Inc., "Splunk," 2014. [Online]. Available: http://www.splunk.com/. [Accessed 13 Feb 2014].

[75] Rapid7, "Metasploit," 2014. [Online]. Available: http://www.metasploit.com/. [Accessed 13 Feb 2014].

[76] D. Kennedy, J. O'Gorman, D. Kearns and M. Aharoni, Metasploit: The Penetration Tester's Guide, No Starch Press, 2011.

[77] Core Security Technologies, "Core Impact Pro," 2014. [Online]. Available: http://www.coresecurity.com/core-impact-pro. [Accessed 13 Feb 2014].

[78] Immunity, Inc., "Canvas," 2004. [Online]. Available: http://immunityinc.com/products-canvas.shtml. [Accessed 13 Feb 2014].

[79] Offensive Security Ltd., "Laki Linux | Rebirth of BackTrack, The Penetration Testing Distribution," 2014. [Online]. Available: http://www.kali.org/. [Accessed 13 Feb 2014].

[80] ThreatTrack Security, Inc., "Sandbox Software & Malware Analysis," 2014. [Online]. Available: http://www.threattracksecurity.com/enterprise-security/sandbox-software.aspx. [Accessed 13 Jan 2014].

[81] FireEye, Inc., "Malware Analysis Tools, Testing & Protection System," 2014. [Online]. Available: http://www.fireeye.com/products-and-solutions/forensic-analysis.html. [Accessed 13 Feb 2014].

[82] Microsoft, "Windows Sysinternals," 29 Jan 2014. [Online]. Available: http://technet.microsoft.com/en-us/sysinternals/default. [Accessed 13 Feb 2014].

[83] Hex-Rays SA, "IDA: About," 25 Jan 2014. [Online]. Available: http://www.hex-rays.com/products/ida/index.shtml. [Accessed 13 Feb 2014].

[84] perl.org, "Perl," 2014. [Online]. Available: http://www.perl.org/. [Accessed 13 Feb 2014].

[85] D. Dougherty and A. Robbins, sed & awk (2nd Edition), O'Reilly Media, 1997.

[86] Wikimedia Foundation, Inc., "Grep," 11 Feb 2014. [Online]. Available: http://en.wikipedia.org/wiki/Grep. [Accessed 13 Feb 2014].

[87] D. Flanagan and Y. Matsumoto, The Ruby Programming Language, O'Reilly, 2008.

[88] D. M. Beazley, Python Essential Reference (4th Edition), Addison-Wesley Professional, 2009.

[89] AccessData Group, LLC, "FTK," 2013. [Online]. Available: http://www.accessdata.com/products/digital-forensics/ftk. [Accessed 13 Feb 2014].

[90]   Guidance Software, Inc., "EnCase Forensic," 2014. [Online]. Available: http://www.guidancesoftware.com/products/Pages/encase-forensic/overview.aspx. [Accessed 13 Feb 2014].

[91]   NIST, "NICE Cybersecurity Workforce Framework," 30 Jul 2013. [Online]. Available: http://csrc.nist.gov/nice/framework/. [Accessed 13 Feb 2014].

[92]   Carnegie Mellon University, "Staffing Your Computer Security Incident Response Team—What Basic Skills Are Needed?," 1 June 2004. [Online]. Available: http://www.cert.org/incident-management/csirt-development/csirt-staffing.cfm. [Accessed 13 Feb 2014].

[93]   R. Bejtlich, "CIRT-Level Response to Advanced Persistent Threat," 3 July 2010. [Online]. Available: http://computer-forensics.sans.org/summit-archives/2010/bejtlich-cirt-level-response.pdf. [Accessed 13 Feb 2014].

[94]   P. Stamp, "Building a World-Class Security Operations Function," Forrester Research, Inc., Cambridge, MA, 2008.

[95]   S. M. Heathfield, "Top Ten Ways to Retain Your Great Employees," 2014. [Online]. Available: http://humanresources.about.com/od/retention/a/more_retention.htm. [Accessed 13 Feb 2014].

[96]   D. J. G. Sujansky and D. J. Ferri-Reed, Keeping the Millennials, Hoboken, New Jersey: John Wiley & Sons, Inc., 2009.

[97]   L. Branham, The 7 Hidden Reasons Employees Leave: How to Recognize the Subtle Signs and Act Before It's Too Late, New York, NY: AMACOM, 2005.

[98]   B. Kaye and S. Jordan-Evans, Love 'Em or Lose 'Em: Getting Good People to Stay, San Francisco, CA: Berrett-Koehler Publishers, Inc., 2008.

[99]   GIAC, "GIAC," 2014. [Online]. Available: http://www.giac.org/. [Accessed 13 Feb 2014].

[100] Offensive Security Ltd., "Offensive Security Training and Professional Services," 2014. [Online]. Available: http://www.offensive-security.com/. [Accessed 13 Feb 2014].

[101] Techweb, "BlackHat Briefings & Training," 2014. [Online]. Available: http://www.blackhat.com/. [Accessed 13 Feb 2014].

[102] DEF CON Communications, Inc., "DEFCON," 2014. [Online]. Available: https://www.defcon.org/. [Accessed 13 Feb 2014].

[103] RSA Conference, "Where the World Talks Security—RSA Conference," 2014. [Online]. Available: http://www.rsaconference.com/. [Accessed 13 Feb 2014].

[104] ToorCon, Inc., "ToorCon," 20 Aug 2013. [Online]. Available: http://toorcon.org/. [Accessed 13 Feb 2014].

[105] ShmooCon, LLC, "ShmooCon," 2014. [Online]. Available: http://www.shmoocon.org/. [Accessed 13 Feb 2014].

[106] Skydogcon, "SkyDogCon News!," 2014. [Online]. Available: http://www.skydogcon.com. [Accessed 13 Feb 2014].

[107] Derbycon, "DerbyCon : Louisville, Kentucky," 2013. [Online]. Available: https://www.derbycon.com/. [Accessed 13 Feb 2014].

[108] BSides, "Security B-Sides," 2014. [Online]. Available: http://www.securitybsides.com/w/page/12194156/FrontPage. [Accessed 13 Feb 2014].

[109] Layerone, "LayerOne," 17 Oct 2013. [Online]. Available: http://www.layerone.org/. [Accessed 13 Feb 2014].

[110] Carnegie Mellon University, "Flocon," 2014. [Online]. Available: http://www.cert.org/flocon/. [Accessed 13 Feb 2014].

[111] Nashville2600.org, "PhreakNIC," 2014. [Online]. Available: http://www.phreaknic.info/. [Accessed 13 Feb 2014].

[112] 2600 Enterprises, Inc., "HOPE X," 2014. [Online]. Available: http://www.hope.net/. [Accessed 13 Feb 2014].

[113] EC-Council, "Hacker Haulted," 2013. [Online]. Available: http://www.hackerhalted.com/. [Accessed 13 Feb 2014].

[114] The SANS Institute, "SANSFIRE 2014," 2014. [Online]. Available: http://www.sans.org/event/sansfire-2014. [Accessed 13 Feb 2014].

[115] USENIX, "USENIX Security Symposium," 2014. [Online]. Available: https://www.usenix.org/conferences/byname/108. [Accessed 13 Feb 2014].

[116] G. Conti, J. Caroland, T. Cook and H. Taylor, "Self-Development for Cyber Warriors," *Small Wars Journal,* 10 Nov 2011.

[117] Microsoft, "Microsoft System Center Configuration Manager 2007 R3," 2014. [Online]. Available: http://www.microsoft.com/en-us/server-cloud/system-center/configuration-manager-features.aspx. [Accessed 13 Feb 2014].

[118] Hewlett-Packard Development Company, L.P., "HP Client Automation software," 2014. [Online]. Available: http://www8.hp.com/us/en/software-solutions/software.html?compURI=1174852. [Accessed 13 Feb 2014].

[119] Symantec Corporation, "Altiris Product Family," 2014. [Online]. Available: http://www.symantec.com/configuration-management. [Accessed 13 Feb 2014].

[120] Microsoft Corporation, "Microsoft Visio 2013 -- flowchart software -- Office.com," 2014. [Online]. Available: http://office.microsoft.com/en-us/visio/. [Accessed 13 Feb 2014].

[121] Lumeta Corporation, "Network Discovery, Leak Detection & Visual Analytics | IPsonar Product Suite from Lumeta," 2014. [Online]. Available: http://www.lumeta.com/product/overview.html. [Accessed 13 Feb 2014].

[122] Wikimedia Foundation, Inc., "traceroute," 4 Feb 2014. [Online]. Available: http://en.wikipedia.org/wiki/Traceroute. [Accessed 13 Feb 2014].

[123] RedSeal Networks, Inc., "RedSeal Networks," 2014. [Online]. Available: http://www.redsealnetworks.com/. [Accessed 13 Feb 2014].

[124] Rapid7, "Vulnerability Management & Risk Management Software | Rapid7," 2014. [Online]. Available: http://www.rapid7.com/products/nexpose/. [Accessed 13 Feb 2014].

[125] Tenable Network Security, "Tenable Nessus," 2014. [Online]. Available: http://www.tenable.com/products/nessus. [Accessed 13 Feb 2014].

[126] BeyondTrust, Inc., "eEye Retina Network Scanner," 2014. [Online]. Available: http://www.beyondtrust.com/Products/RetinaNetworkSecurityScanner/. [Accessed 13 Feb 2014].

[127] K. Katterjohn, "Port Scanning Techniques," Packetstorm, 8 March 2007. [Online]. Available: http://packetstormsecurity.org/files/view/54973/port-scanning-techniques.txt. [Accessed 13 Feb 2014].

[128] G. Lyon, "Nmap," 2014. [Online]. Available: http://nmap.org/. [Accessed 13 Feb 2014].

[129] G. F. Lyon, Nmap Network Scanning: The Official Nmap Project Guide to Network Discovery and Security Scanning, Sunnyvale: Insecure.com LLC, 2009.

[130] M. Zalewski, "p0f v3," 2012. [Online]. Available: http://lcamtuf.coredump.cx/p0f3/. [Accessed 13 Feb 2014].

[131] Geeknet, Inc., "X probe - active OS fingerprinting tool," 4 Jun 2013. [Online]. Available: http://sourceforge.net/projects/xprobe/. [Accessed 13 Feb 2014].

[132] Cisco, "Intelligent Cybersecurity Solutions | Sourcefire," 2014. [Online]. Available: http://www.sourcefire.com/security-technologies/cyber-security-products/3d-system/network-awareness. [Accessed 13 Feb 2014].

[133] Ettercap Project, "Ettercap Home Page," 2014. [Online]. Available: http://ettercap.github.io/ettercap/. [Accessed 13 Feb 2014].

[134] J. M. Allen, "OS and Application Fingerprinting Techniques," 22 Sept 2007. [Online]. Available: http://www.sans.org/reading_room/whitepapers/protocols/os-application-fingerprinting-techniques_1891. [Accessed 13 Feb 2014].

[135] K. Scarfone and P. Mell, "NIST Special Publication 800-94: Guide to Intrusion Detection and Prevention Systems (IDPS)," Feb 2007. [Online]. Available: http://csrc.nist.gov/publications/nistpubs/800-94/SP800-94.pdf. [Accessed 13 Feb 2014].

[136] N. Pappas, "Network IDS & IPS Deployment Strategies," 2 April 2008. [Online]. Available: http://www.sans.org/reading_room/whitepapers/detection/network-ids-ips-deployment-strategies_2143. [Accessed 13 Feb 2014].

[137] M. Heckathorn, "Network Monitoring for Web-Based Threats," Feb 2011. [Online]. Available: http://www.sei.cmu.edu/reports/11tr005.pdf. [Accessed 13 Feb 2014].

[138] Cisco, "Cisco IOS NetFlow," 2014. [Online]. Available: http://www.cisco.com/en/US/products/ps6601/products_ios_protocol_group_home.html. [Accessed 13 Feb 2014].

[139] B. Claise, "Request for Comments: 3954 Cisco Systems NetFlow Services Export Version 9," Oct 2004. [Online]. Available: http://www.ietf.org/rfc/rfc3954.txt. [Accessed 13 Feb 2014].

[140] Wikimedia Foundation, Inc., "Pen register," 8 Feb 2014. [Online]. Available: http://en.wikipedia.org/wiki/Pen_register. [Accessed 13 Feb 2014].

[141] B. Visscher, "Sguil," May 2011. [Online]. Available: http://sguil.sourceforge.net/. [Accessed 13 Feb 2014].

[142] M. Lucas, Network Flow Analysis, No Starch Press, 2010.

[143] International Business Machines, "IBM - Proventia Network Intrusion Prevention System," 2014. [Online]. Available: http://www-935.ibm.com/services/in/en/it-services/proventia-network-intrusion-prevention-system.html. [Accessed 13 Feb 2014].

[144] C. L. Van Jacobson and S. McCanne, "Tcpdump/Libpcap public repository," 2014. [Online]. Available: http://www.tcpdump.org/. [Accessed 13 Feb 2014].

[145] Wikimedia Foundation, Inc., "Optical Carrier Transmission Rates," 3 Feb 2014. [Online]. Available: http://en.wikipedia.org/wiki/Optical_Carrier_transmission_rates. [Accessed 13 Feb 2014].

[146] Emulex Corporation, "DAG Packet Capture Cards," 2014. [Online]. Available: http://www.emulex.com/products/network-visibility-products/dag-packet-capture-cards/features/. [Accessed 13 Feb 2014].

[147] EMC Corporation, "RSA NetWitness," 2014. [Online]. Available: http://www.emc.com/security/rsa-netwitness.htm. [Accessed 14 Feb 2014].

[148] Solera Networks, Inc., "Appliances | Solera Networks," 2014. [Online]. Available: http://www.soleranetworks.com/products/security-analytics/appliances/. [Accessed 14 Feb 2014].

[149] AccessData Group, LLC, "SilentRunner Sentinel Network Forensics Software," 2014. [Online]. Available: http://www.accessdata.com/products/cyber-security/silent-runner. [Accessed 14 Feb 2014].

[150] Oracle Corporation, "Oracle Solaris ZFS Administration Guide," 2010. [Online]. Available: http://docs.oracle.com/cd/E19253-01/819-5461/index.html. [Accessed 13 Feb 2014].

[151] J. Østergaard and E. Bueso, "The Software-RAID HOWTO," 6 March 2010. [Online]. Available: http://tldp.org/HOWTO/Software-RAID-HOWTO.html. [Accessed 14 Feb 2014].

[152] NVIDIA Corporation, "CUDA," 2014. [Online]. Available: http://www.nvidia.com/object/cuda_home_new.html. [Accessed 14 Feb 2014].

[153] Khronos Group, "OpenCL," 2014. [Online]. Available: http://www.khronos.org/opencl/. [Accessed 14 Feb 2014].

[154] CSP, Inc., "Myricom Cybersecurity," 2014. [Online]. Available: https://www.myricom.com/solutions/cybersecurity.html. [Accessed 14 Feb 2014].

[155] VMware, Inc., "VMware ESX and VMware ESXi," 2009. [Online]. Available: http://www.vmware.com/files/pdf/VMware-ESX-and-VMware-ESXi-DS-EN.pdf. [Accessed 14 Feb 2014].

[156] Citrix Systems, "The Xen Project," 2014. [Online]. Available: http://www.xenproject.org/. [Accessed 14 Feb 2014].

[157] E. Valente, "Capturing 10G versus 1G Traffic Using Correct Settings!," 2009. [Online]. Available: http://www.sans.org/reading_room/whitepapers/detection/capturing-10g-1g-traffic-correct-settings_33043. [Accessed 14 Feb 2014].

[158] Norman Shark, "Malware Analysis - Normak Shark malware analysis sandbox," 2014. [Online]. Available: http://normanshark.com/products-solutions/products/malware-analysis-mag2/. [Accessed 14 Feb 2014].

[159] Damballa, Inc., "Next Generation Anti-Virus: The Pros and Cons of Dynamic Malware Dissection," 1 Aug 2011. [Online]. Available: http://www.damballa.com/downloads/r_pubs/WP_Next_Generation_Anti-Virus.pdf. [Accessed 14 Feb 2014].

[160] L. R. Even, "Intrusion Detection FAQ: What is a Honeypot?," 12 July 2000. [Online]. Available: http://www.sans.org/security-resources/idfaq/honeypot3.php. [Accessed 14 Feb 2014].

[161] The Honeynet Project, Know Your Enemy: Learning about Security Threats (2nd Edition), Addison-Wesley Professional, 2004.

[162] Gartner, Inc., "Hype Cycle for Information Security, 2003," Gartner, Inc., Stamford, CT, 2003.

[163] Wikimedia Foundation, Inc., "Cryptographic Hash Function," 12 Feb 2014. [Online]. Available: http://en.wikipedia.org/wiki/Cryptographic_hash_function. [Accessed 14 Feb 2014].

[164] Wikimedia Foundation, Inc., "Ring (computer security)," 14 Feb 2014. [Online]. Available: http://en.wikipedia.org/wiki/Ring_%28computer_security%29. [Accessed 14 Feb 2014].

[165] J. Rutkowska, "Introducing Blue Pill," 22 June 2006. [Online]. Available: http://the-invisiblethings.blogspot.com/2006/06/introducing-blue-pill.html. [Accessed 14 Feb 2014].

[166] N. L. Petroni, Jr., T. Fraser, J. Molina and W. A. Arbaugh, "Copilot - a Coprocessor-based Kernel Runtime Integrity Monitor," in *USENIX Security 2004*, San Diego, CA, 2004.

[167] Trusted Computing Group, "Trusted Platform Module," 2014. [Online]. Available: http://www.trustedcomputinggroup.org/developers/trusted_platform_module/. [Accessed 14 Feb 2014].

[168] G. Hoglund and J. Butler, Rootkits: Subverting the Windows Kernel, Addison-Wesley Professional, 2005.

[169] G. Ollmann, "Enterprise Protection Against Botnet Breaches," 2009. [Online]. Available: http://www.damballa.com/downloads/r_pubs/WP_SerialVariantEvasionTactics.pdf. [Accessed 14 Feb 2014].

[170] Mandiant Corporation, "M-Trends 2010," January 2010. [Online]. Available: https://www.mandiant.com/resources/mandiant-reports/. [Accessed 14 Feb 2014].

[171] D. Raywood, "Antivirus programs only detect 18% of zero-day malware," 11 Aug 2010. [Online]. Available: http://www.scmagazine.com.au/News/224259,antivirus-programs-only-detect-18-of-zero-day-malware.aspx. [Accessed 14 Feb 2014].

[172] D. Goodin, "Anti-virus protection gets worse," 21 Dec 2007. [Online]. Available: http://www.channelregister.co.uk/2007/12/21/dwindling_antivirus_protection/. [Accessed 14 Feb 2014].

[173] R. McMillan, "Is Antivirus Software a Waste of Money?," 2 Mar 2012. [Online]. Available: http://www.wired.com/wiredenterprise/2012/03/antivirus/. [Accessed 14 Feb 2014].

[174] AV-Comparatives.org, "Welcome to AV-Comparatives.org," 2014. [Online]. Available: http://www.av-comparatives.org/. [Accessed 14 Feb 2014].

[175] IBM, "IBM Security Server Protection," 2014. [Online]. Available: http://www-01.ibm.com/software/tivoli/products/security-server-protection/. [Accessed 14 Feb 2014].

[176] McAfee, Inc., "McAfee Host Intrusion Prevention for Server," 2014. [Online]. Available: http://www.mcafee.com/us/products/host-ips-for-server.aspx. [Accessed 14 Feb 2014].

[177] Symantec Corporation, "Symantec Endpoint Protection," 2014. [Online]. Available: http://www.symantec.com/endpoint-protection. [Accessed 14 Feb 2014].

[178] Sophos, Ltd., "Enterprise Endpoint Antivirus Suite | Data Protection for All Devices," 2014. [Online]. Available: http://www.sophos.com/en-us/products/enduser-protection-suites.aspx. [Accessed 14 Feb 2014].

[179] TechTarget, "Application Blacklisting," Jun 2011. [Online]. Available: http://searchsecurity.techtarget.com/definition/application-blacklisting. [Accessed 14 Feb 2014].

[180] Microsoft, "Windows 7 AppLocker Executive Overview," 4 Dec 2013. [Online]. Available: http://technet.microsoft.com/en-us/library/dd548340%28v=ws.10%29.aspx. [Accessed 14 Feb 2014].

[181] Bit9, Inc., "Application Whitelisting," 2014. [Online]. Available: https://www.bit9.com/solutions/application-whitelisting/. [Accessed 14 Feb 2014].

[182] McAfee, Inc., "McAfee Application Control," 2014. [Online]. Available: http://www.mcafee.com/us/products/application-control.aspx. [Accessed 14 Feb 2014].

[183] C. Shaffer and C. Cuevas, "Raising the White Flag: Bypassing Application White Listing," 28 Jan 2012. [Online]. Available: http://blog.c22.cc/2012/01/28/shmoocon-2012-raising-the-white-flag/. [Accessed 14 Feb 2014].

[184] A. Beuhring and K. Salous, "ShmooCon 2014 - Raising Costs for Your Attackers Instead of Your CFO," Jan 2014. [Online]. Available: https://archive.org/details/ShmooCon2014_Raising_Costs_for_Your_Attackers_Instead_of_Your_CFO. [Accessed 14 Feb 2014].

[185] Geeknet, Inc., "Open Source Tripwire," 5 Aug 2013. [Online]. Available: http://sourceforge.net/projects/tripwire/. [Accessed 14 Feb 2014].

[186] Tripwire, Inc., "Tripwire," 2014. [Online]. Available: http://www.tripwire.com/. [Accessed 14 Feb 2014].

[187] Inverse, "Packetfence," 2013. [Online]. Available: http://www.packetfence.org/home. html. [Accessed 14 Feb 2014].

[188] McAfee, Inc., "Network Security | McAfee Products," 2014. [Online]. Available: http://www.mcafee.com/us/products/network-access-control.aspx. [Accessed 14 Feb 2014].

[189] Juniper Networks, Inc., "Unified Access Control," 2014. [Online]. Available: http://www.juniper.net/us/en/products-services/security/uac/. [Accessed 14 Feb 2014].

[190] Microsoft, "Network Policy and Access Services," 2014. [Online]. Available: http://technet.microsoft.com/en-us/network/bb545879.aspx. [Accessed 14 Feb 2014].

[191] S. Wilkins, "Switchport Security Concepts and Configurations," 1 July 2011. [Online]. Available: http://www.informit.com/articles/article.aspx?p=1722561. [Accessed 14 Feb 2014].

[192] J. Guttman, A. Herzog, J. Millen, L. Monk, J. Ramsdell, J. Sheehy, B. Sniffen, G. Coker and P. Loscocco, "Attestation: Evidence and Trust," March 2007. [Online]. Available: http://www.mitre.org/publications/technical-papers/attestation-evidence-and-trust. [Accessed 14 Feb 2014].

[193] P. N. Ayuso, "The netfilter.org project," 2010. [Online]. Available: http://www.netfilter.org/. [Accessed 14 Feb 2014].

[194] The OpenBSD Team, "PF: The OpenBSD Packet Filter," 13 Jan 2014. [Online]. Available: http://www.openbsd.org/faq/pf/. [Accessed 14 Feb 2014].

[195] Symantec Corporation, "Symantec Endpoint Protection," 2014. [Online]. Available: http://www.symantec.com/endpoint-protection. [Accessed 14 Feb 2014].

[196] Check Point Software Technologies Ltd., "Firewall & Compliance Check," 2014. [Online]. Available: http://www.checkpoint.com/products/firewall-compliance-check/index.html. [Accessed 14 Feb 2014].

[197] McAfee, Inc., "McAfee Host Intrusion Prevention for Desktop," 2014. [Online]. Available: http://www.mcafee.com/us/products/host-ips-for-desktop.aspx. [Accessed 14 Feb 2014].

[198] Microsoft, "Group Policy for Beginners," 27 April 2011. [Online]. Available: http://technet.microsoft.com/en-us/library/hh147307%28WS.10%29.aspx. [Accessed 14 Feb 2014].

[199] Raytheon Company, "Raytheon SureView," 2014. [Online]. Available: http://www.raytheon.com/capabilities/products/cybersecurity/insiderthreat/products/sureview/. [Accessed 14 Feb 2014].

[200] ObserveIT, "ObserveIT," 2014. [Online]. Available: http://www.observeit.com/. [Accessed 14 Feb 2014].

[201] Mandiant, LLC, "Mandiant Intelligent Response," 2014. [Online]. Available: https://www.mandiant.com/products/mandiant-platform/intelligent-response. [Accessed 13 Jan 2014].

[202] AccessData, "AccessData CIRT," 2013. [Online]. Available: http://accessdata.com/products/cyber-security-incident-response/cirt. [Accessed 14 Feb 2014].

[203] HBGary, Inc., "Responder Pro | HBGary," 2013. [Online]. Available: http://www.hbgary.com/products/responder_pro. [Accessed 14 Feb 2014].

[204] Guidance Software, Inc., "Encase Enterprise," 2014. [Online]. Available: http://www.guidancesoftware.com/products/Pages/encase-enterprise/overview.aspx. [Accessed 14 Feb 2014].

[205] B. M. Posey, "Demystifying the 'Blue Screen of Death'," 2014. [Online]. Available: http://technet.microsoft.com/en-us/library/cc750081.aspx. [Accessed 14 Feb 2014].

[206] D. Haye, "Harness the power of SIEM," 15 April 2009. [Online]. Available: http://www.sans.org/reading_room/whitepapers/detection/harness-power-siem_33204. [Accessed 14 Feb 2014].

[207] J. Voorhees, "Distilling Data in a SIM: A Strategy for the Analysis of Events in the ArcSight ESM," 26 Sept 2007. [Online]. Available: http://www.sans.org/reading_room/whitepapers/detection/distilling-data-sim-strategy-analysis-events-arcsight-esm_1916. [Accessed 14 Feb 2014].

[208] A. A. Chuvakin, "Anton Chuvakin Homepage," 12 Nov 2013. [Online]. Available: http://www.chuvakin.org/. [Accessed 14 Feb 2014].

[209] R. Marty, "Security Intelligence and Big Data | raffy.ch," 19 Jan 2014. [Online]. Available: http://raffy.ch/blog/. [Accessed 14 Feb 2014].

[210] R. DeStefano, "visiblerisk blog," 2014. [Online]. Available: http://www.visiblerisk.com/blog/. [Accessed 14 Feb 2014].

[211] K. Kent and M. Souppaya, "NIST Special Publication 800-92: Guide to Computer Security Log Management," Sept 2006. [Online]. Available: http://csrc.nist.gov/publications/nistpubs/800-92/SP800-92.pdf. [Accessed 13 Feb 2014].

[212] A. A. Chuvakin and K. J. Schmidt, Logging and Log Management: The Authoritative Guide to Dealing with Syslog, Audit Logs, Events, Alerts and other IT 'Noise', Boston, MA: Syngress, 2012.

[213] M. Nicolett and K. M. Kavanagh, "Critical Capabilities for Security Information and Event Management," 21 May 2012. [Online]. Available: https://www.gartner.com/doc/2022315. [Accessed 14 Feb 2014].

[214] ISACA, "Security Information and Event Management: Business Benefits and Security, Governance and Assurance Perspective," 28 Dec 2010. [Online]. Available: http://www.isaca.org/Knowledge-Center/Research/ResearchDeliverables/Pages/Security-Information-and-Event-Management-Business-Benefits-and-Security-Governance-and-Assurance-Perspective.aspx. [Accessed 14 Feb 2014].

[215] UBM TechWeb, "Security Monitoring : Tech Center," 2014. [Online]. Available: http://www.darkreading.com/monitoring/. [Accessed 14 Feb 2014].

[216] M. Nicolett and K. M. Kavanagh, "Magic Quadrant for Security Information and Event Management," 7 May 2013. [Online]. Available: https://www.gartner.com/doc/2477018/magic-quadrant-security-information-event. [Accessed 14 Feb 2014].

[217] J. Dean and S. Ghemawat, Dec 2004. [Online]. Available: http://research.google.com/archive/mapreduce.html. [Accessed 14 Feb 2014].

[218] Splunk Inc., "Splunk 4 Down Under," 27 Aug 2009. [Online]. Available: http://blogs.splunk.com/2009/08/27/splunk-4-down-under/. [Accessed 14 Feb 2014].

[219] A. Chuvakin, "Something Fun About Using SIEM," 18 Feb 2011. [Online]. Available: http://www.slideshare.net/anton_chuvakin/something-fun-about-using-siem-by-dr-anton-chuvakin. [Accessed 14 Feb 2014].

[220] www.decision-making-confidence.com, "Kepner Tregoe Decision Making The Steps, The Pros and The Cons," 2013. [Online]. Available: http://www.decision-making-confidence.com/kepner-tregoe-decision-making.html. [Accessed 14 Feb 2014].

[221] AlienVault, "OSSIM: Open Source SIEM & Open Threat Exchange Projects," 2014. [Online]. Available: http://www.alienvault.com/open-threat-exchange/projects. [Accessed 13 Feb 2014].

[222] J. Sissel, "logstash," 2013. [Online]. Available: http://logstash.net/. [Accessed 14 Feb 2014].

[223] R. Khan, "Kibana. Make sense of a mountain of logs," 2014. [Online]. Available: http://kibana.org/. [Accessed 14 Feb 2014].

[224] B. Tung, "Common Intrusion Detection Framework (CIDF)," 10 Sept 1999. [Online]. Available: http://gost.isi.edu/cidf/. [Accessed 14 Feb 2014].

[225] R. Danyliw, J. Meijer and Y. Demchenko, "The Incident Object Description Exchange Format," Dec 2007. [Online]. Available: http://www.ietf.org/rfc/rfc5070.txt. [Accessed 14 Feb 2014].

[226] BusinessWire, "ICSA Labs IDS Consortium Announces Network Intrusion Detection System Alert Specification Format," 23 Feb 2004. [Online]. Available: http://www.businesswire.com/news/home/20040223005073/en/ICSA-Labs-IDS-Consortium-Announces-Network-Intrusion. [Accessed 14 Feb 2014].

[227] ISACA, "WebTrends Enhanced Log Format," 17 Jul 2008. [Online]. Available: http://download.logreport.org/pub/current/doc/user-manual/ch10s05.html. [Accessed 14 Feb 2014].

[228] International Business Machines, "Common Event Infrastructure," 2005. [Online]. Available: http://www-01.ibm.com/software/tivoli/features/cei/. [Accessed 14 Feb 2014].

[229] ArcSight, Inc., "Common Event Format," 17 July 2009. [Online]. Available: http://mita-tac.wikispaces.com/file/view/CEF+White+Paper+071709.pdf. [Accessed 14 Feb 2014].

[230] The MITRE Corporation, "Common Event Expression (CEE)," 30 May 2013. [Online]. Available: http://cee.mitre.org/. [Accessed 14 Feb 2014].

[231] Cornell University Law School, "Federal Rules of Evidence," 1 Dec 2013. [Online]. Available: http://www.law.cornell.edu/rules/fre. [Accessed 14 Feb 2014].

[232] D. Levin, "Logs & The Law: What is Admissible in Court?," 12 June 2006. [Online]. Available: http://www.slideshare.net/loglogic/logs-the-law-what-is-admissible-in-court. [Accessed 14 Feb 2014].

[233] Wikimedia Foundation, Inc., "Cloud Computing," 14 Feb 2014. [Online]. Available: http://en.wikipedia.org/wiki/Cloud_computing. [Accessed 14 Feb 2014].

[234] OSSEC, "Log samples," 26 June 2010. [Online]. Available: http://www.ossec.net/doc/log_samples/. [Accessed 13 Feb 2014].

[235] K. Gorzelak, T. Grudziecki, P. Jacewicz, P. Jaroszewski, Ł. Juszczyk and P. Kijewski, "Proactive detection of network security incidents," 7 Dec 2011. [Online]. Available: http://www.enisa.europa.eu/activities/cert/support/proactive-detection. [Accessed 14 Feb 2014].

[236] Net Optics, Inc., "NetOptics," 2014. [Online]. Available: http://netoptics.com/. [Accessed 14 Feb 2014].

[237] Fluke Corporation, "Tap Solutions," 2014. [Online]. Available: http://www.flukenetworks.com/enterprise-network/network-monitoring/Tap-Solutions. [Accessed 14 Feb 2014].

[238] Network Critical Solutions Limited, "Network Critical," 2011. [Online]. Available: http://www.networkcritical.com/. [Accessed 14 Feb 2014].

[239] Datacom Systems Inc., "Datacom Systems Inc. Network Taps," 2013. [Online]. Available: http://www.datacomsystems.com/products/network-taps.asp. [Accessed 14 Feb 2014].

[240] D. G. Gomez, "Receive-only UTP cables and Network Taps," May 2006. [Online]. Available: http://dgonzalez.net/pub/roc/roc.pdf. [Accessed 14 Feb 2014].

[241] Owl Computing Technologies, Inc., "Owl Computing Technologies, Inc.," 2014. [Online]. Available: http://www.owlcti.com/. [Accessed 14 Feb 2014].

[242] Hewlett Packard Development Company, L.P., "Tenix Data Diode Based Solutions from HP," 19 April 2004. [Online]. Available: ftp://ftp.hp.com/pub/services/security/products/info/tenix_data_wp.pdf. [Accessed 14 Feb 2014].

[243] M. W. Stevens, "An Implementation of an Optical Data Diode," 14 July 1999. [Online]. Available: http://www.dsto.defence.gov.au/publications/2110/DSTO-TR-0785.pdf. [Accessed 14 Feb 2014].

[244] M. Trojnara, "stunnel: stunnel - multiplatform SSL tunneling proxy," 26 Nov 2013. [Online]. Available: http://www.stunnel.org/. [Accessed 14 Feb 2014].

[245] Wikimedia Foundation, Inc., "VLAN Hopping," 18 Dec 2013. [Online]. Available: http://en.wikipedia.org/wiki/VLAN_hopping. [Accessed 14 Feb 2014].

[246] R. Bejtlich, "What Is APT and What Does It Want?," 16 Jan 2010. [Online]. Available: http://taosecurity.blogspot.com/2010/01/what-is-apt-and-what-does-it-want.html. [Accessed 4 Feb 2014].

[247] D. Edwards, "What is DevOps?," 23 Feb 2012. [Online]. Available: http://dev2ops.
org/2010/02/what-is-devops/. [Accessed 13 Jan 2014].

[248] FireEye, Inc., "FireEye," 2014. [Online]. Available: http://www.fireeye.com/.
[Accessed 13 Jan 2014].

[249] C. Guarnieri, "Automated Malware Analysis - Cuckoo Sandbox," 2013. [Online].
Available: http://www.cuckoosandbox.org/. [Accessed 13 Jan 2014].

[250] VMware, Inc., "VMware Workstation," 2014. [Online]. Available: http://www.
vmware.com/products/workstation/. [Accessed 13 Jan 2014].

[251] Microsoft, "Microsoft Windows," 2014. [Online]. Available: http://windows.microsoft.
com/en-us/windows/home. [Accessed 13 Jan 2014].

[252] Microsoft, "Windows Sysinternals," 2014. [Online]. Available: http://technet.micro-
soft.com/en-us/sysinternals/default. [Accessed 13 Jan 2014].

[253] T. Hungenberg and M. Eckert, "INetSim: Internet Services Simulation Suite," 2013.
[Online]. Available: http://www.inetsim.org/. [Accessed 13 Jan 2014].

[254] W. Shields, "An Introduction to chopshop," 7 Nov 2012. [Online]. Available: http://
www.mitre.org/capabilities/cybersecurity/overview/cybersecurity-blog/an-introduc-
tion-to-chopshop-network-protocol. [Accessed 13 Jan 2014].

[255] Hex-Rays SA, "IDA: About," 25 Jul 2012. [Online]. Available: http://www.hex-rays.
com/products/ida/index.shtml. [Accessed 13 Jan 2014].

[256] Microsoft, "Windows Driver Kit (WDK) and Debugging Tools for Windows (WinDbg)
downloads," 2014. [Online]. Available: http://www.microsoft.com/whdc/devtools/
debugging/default.mspx. [Accessed 13 Jan 2014].

[257] O. Yuschuk, "OllyDbg," 2013. [Online]. Available: http://www.ollydbg.de/. [Accessed
13 Jan 2014].

[258] Guidance Software, Inc., "EnCase Forensic," 2014. [Online]. Available: http://www.
guidancesoftware.com/products/Pages/encase-forensic/overview.aspx. [Accessed 13
Jan 2014].

[259] AccessData Group, LLC, "FTK," 2013. [Online]. Available: http://www.accessdata.
com/products/digital-forensics/ftk. [Accessed 13 Jan 2014].

[260] B. Carrier, "The Sleuth Kit (TSK) & Autopsy: Open Source Digital Forensics Tools,"
2013. [Online]. Available: http://www.sleuthkit.org/. [Accessed 13 Jan 2014].

[261] SANS Institute, "SANS SIFT Kit/Workstation: Investigative Forensic Toolkit
Download," 2014. [Online]. Available: http://computer-forensics.sans.org/community/
downloads. [Accessed 13 Jan 2014].

[262] M. Ligh, A. Case, J. Levy and A. Walters, "volatility - An advanced memory forensics framework," 2014. [Online]. Available: https://code.google.com/p/volatility/.
[Accessed 13 Jan 2014].

[263] Guidance Software, Inc., "Tableau Forensic Products - Forensic Bridges," 2014.
[Online]. Available: http://www.tableau.com/index.php?pageid=products&category=
forensic_bridges. [Accessed 13 Jan 2014].

[264] Digital Intelligence, "FRED," 2014. [Online]. Available: http://www.digitalintelligence.com/products/fred/. [Accessed 13 Jan 2014].

[265] X. Kovah, "Open Security Training .Info," 2014. [Online]. Available: http://opensecuritytraining.info/. [Accessed 13 Jan 2014].

[266] Wikimedia Foundation, Inc., "Wiki," 12 Feb 2012. [Online]. Available: http://en.wikipedia.org/wiki/Wiki. [Accessed 14 Feb 2012].

[267] Palantir Technologies, "Home | Palantir," 2013. [Online]. Available: https://www.palantir.com/. [Accessed 13 Jan 2014].

[268] FS-ISAC, "Welcome to the Cyber Intelligence Repository Landing Page," 2013. [Online]. Available: https://www.fsisac.com/CyberIntelligenceRepository. [Accessed 4 Feb 2014].

[269] REN-ISAC, "collective-intelligence-framework," 2014. [Online]. Available: https://code.google.com/p/collective-intelligence-framework/. [Accessed 4 Feb 2014].

[270] Lockheed Martin Corporation, "Cyber Intelligence Enteprise Solutions," 2013. [Online]. Available: http://www.lockheedmartin.com/us/what-we-do/information-technology/cyber-security/cyber-intelligence-enterprise.html. [Accessed 4 Feb 2014].

[271] Cyber Squared Inc., "ThreatConnect | Threat Analysis and Community," 2014. [Online]. Available: http://www.threatconnect.com/. [Accessed 4 Feb 2014].

[272] The MITRE Corporation, "Collaborative Research into Threats (CRITs) | The MITRE Corporation," 2014. [Online]. Available: http://www.mitre.org/research/technology-transfer/technology-licensing/collaborative-research-into-threats-crits. [Accessed 13 Jan 2014].

[273] HBGary, Inc., "Responder Pro | HBGary," 2013. [Online]. Available: http://www.hbgary.com/products/responder_pro. [Accessed 13 Jan 2014].

[274] Guidance Software, Inc., "Encase Enterprise," 2014. [Online]. Available: http://www.guidancesoftware.com/products/Pages/encase-enterprise/overview.aspx. [Accessed 13 Jan 2014].

[275] FIRST.org, Inc., "FIRST," 2014. [Online]. Available: http://www.first.org/. [Accessed 14 Feb 2014].

[276] ISAC Council, "National Council of ISACs," 2014. [Online]. Available: http://www.isaccouncil.org/. [Accessed 16 Feb 2014].

[277] B. Bakis, "Blueprint for Cyber Threat Sharing - Lessons Learned & Challenges," 13 Dec 2013. [Online]. Available: http://www.mitre.org/capabilities/cybersecurity/overview/cybersecurity-blog/blueprint-for-cyber-threat-sharing-lessons. [Accessed 16 Feb 2014].

[278] Bit9, Inc., "Simplify and Speed Cyber Forensics Investigations," 2014. [Online]. Available: https://www.bit9.com/solutions/cloud-services/cyber-forensics/. [Accessed 16 Feb 2014].

[279] CrowdStrike, Inc., "CrowdStrike: a Stealth-Mode Security Start-Up," 2014. [Online]. Available: http://www.crowdstrike.com/. [Accessed 16 Feb 2014].

[280] VeriSign, Inc., "iDefense Security Intelligence Services," 2014. [Online]. Available: http://www.verisigninc.com/en_US/cyber-security/index.xhtml. [Accessed 16 Feb 2014].

[281] Symantec Corporation, "DeepSight Security Intelligence Products," 2014. [Online]. Available: http://www.symantec.com/deepsight-products. [Accessed 16 Feb 2014].

[282] The MITRE Corporation, "Cyber Intelligence Threat Analysis," 5 Jul 2013. [Online]. Available: http://msm.mitre.org/directory/areas/threatanalysis.html. [Accessed 16 Feb 2014].

[283] Wikimedia Foundation, Inc., "Fast flux," 15 Jun 2013. [Online]. Available: http://en.wikipedia.org/wiki/Fast_flux. [Accessed 14 Feb 2014].

[284] Best Practical Solutions LLC, "RTIR: RT for Incident Response," 2014. [Online]. Available: http://bestpractical.com/rtir/. [Accessed 16 Feb 2014].

[285] Wikimedia Foundation, Inc., "Wiki," 1 Feb 2014. [Online]. Available: http://en.wikipedia.org/wiki/Wiki. [Accessed 16 Feb 2014].

[286] J. Viega, "Why most companies shouldn't run intrusion prevention," 4 Dec 2008. [Online]. Available: http://broadcast.oreilly.com/2008/12/why-most-companies-shouldnt-ru.html. [Accessed 16 Feb 2014].

[287] J. Allen, D. Gabbard and C. May, "Outsourcing Managed Security Services," January 2003. [Online]. Available: http://resources.sei.cmu.edu/library/asset-view.cfm?assetID=6319. [Accessed 16 Feb 2014].

[288] Carnegie Mellon University CERT, "Creating a Computer Security Incident Response Team: A Process for Getting Started," 26 Feb 2006. [Online]. Available: http://www.cert.org/incident-management/products-services/creating-a-csirt.cfm. [Accessed 16 Feb 2014].

[289] A. Jaquith, Security Metrics: Replacing Fear, Uncertainty, and Doubt, Boston, MA: Pearson Education, Inc., 2007.

[290] Hewlett-Packard Development Company, L.P., "State of security operations," Jan 2014. [Online]. Available: http://www.hp.com/go/StateOfSecOps. [Accessed 3 Mar 2014].

[291] National Institute of Standards and Technology (NIST), "Managing Information Security Risk: Organization, Mission, and Information System View," Gaithersburg, MD, 2011.

[292] Risk Steering Committee, "DHS Risk Lexicon," Sept 2010. [Online]. Available: http://www.dhs.gov/dhs-risk-lexicon. [Accessed 16 Feb 2014].

[293] M. Cloppert, "Security Intelligence: Attacking the Kill Chain," SANS, 14 Oct 2009. [Online]. Available: http://computer-forensics.sans.org/blog/2009/10/14/security-intelligence-attacking-the-kill-chain/. [Accessed 13 Feb 2014].

[294] The MITRE Corporation, "Cyber Observable eXpression – CybOX™," 6 Feb 2012. [Online]. Available: http://makingsecuritymeasurable.mitre.org/docs/cybox-intro-handout.pdf. [Accessed 16 Feb 2014].

[295] National Security Council, "Cybersecurity," 29 May 2009. [Online]. Available: http://www.whitehouse.gov/issues/foreign-policy/cybersecurity. [Accessed 16 Feb 2014].

[296] IGnet, "Council of the Inspectors General on Integrity and Efficiency," 2014. [Online]. Available: http://www.ignet.gov/. [Accessed 16 Feb 2014].

[297] Chairman of the Joint Chiefs of Staff, "Chairman of the Joint Chiefs of Staff Instruction (CJCSI) 5120.02," 30 Nov 2004. [Online]. Available: http://www.bits.de/NRANEU/others/jp-doctrine/cjcsi5120_02%2804%29.pdf. [Accessed 16 Feb 2014].

[298] Wikimedia Foundation, Inc., "ZFS," 13 Feb 2012. [Online]. Available: http://en.wikipedia.org/wiki/ZFS. [Accessed 13 Feb 2012].

[299] Wikimedia Foundation, Inc., "Whac-A-Mole," 27 Jan 2012. [Online]. Available: http://en.wikipedia.org/wiki/Whac-A-Mole. [Accessed 20 Feb 2012].

[300] Sensage, Inc., "Sensage," 2011. [Online]. Available: http://www.sensage.com/. [Accessed 24 Feb 2012].

[301] J. Seitz, Gray Hat Python: Python Programming for Hackers and Reverse Engineers, San Francisco: No Starch Press, 2009.

[302] SANS Institute, "Cyber Attack Threat Map," 2008. [Online]. Available: http://www.sans.org/whatworks/poster_fall_08.pdf. [Accessed 12 Feb 2012].

[303] Nitrosecurity, "Enterprise SIEM - NitroSecurity - Security Information & Event Management," 2012. [Online]. Available: http://www.nitrosecurity.com/products/enterprise-security-manager/. [Accessed 5 Apr 2012].

[304] B. Keyes, "The Drake Equation," 2005. [Online]. Available: http://www.activemind.com/Mysterious/Topics/seti/drake_equation.html. [Accessed 14 Feb 2012].

[305] Hewlett-Packard Development Company, L.P., "ArcSight Interactive Discovery," November 2011. [Online]. Available: http://www.arcsight.com/collateral/ArcSight_Interactive_Discovery.pdf. [Accessed 24 Feb 2012].

[306] E. Fitzgerald, "Windows Security Logging and Other Esoterica," 2011. [Online]. Available: http://blogs.msdn.com/b/ericfitz/. [Accessed 5 Mar 2012].

[307] Endace Limited, "Select your DAG card," 2011. [Online]. Available: http://www.endace.com/compare-dag-card-models.html. [Accessed 13 Feb 2012].

[308] EC-Council, "Certified Ethical Hacker," 201. [Online]. Available: http://www.eccouncil.org/courses/certified_ethical_hacker.aspx. [Accessed 14 Feb 2012].

[309] R. Dawes, "OWASP WebScarab Project," 18 Jan 2012. [Online]. Available: https://www.owasp.org/index.php/Category:OWASP_WebScarab_Project. [Accessed 13 Feb 2012].

[310] Carnegie Mellon University, "Capability Maturity Model Integration (CMMI)," 2012. [Online]. Available: http://www.sei.cmu.edu/cmmi/. [Accessed 14 Feb 2012].

[311] BackTrack Linux, "BackTrack Linux – Penetration Testing Distribution," 2012. [Online]. Available: http://www.backtrack-linux.org/. [Accessed 13 Feb 2012].

[312] ArcSight, LLC, "Common Event Format," 2012. [Online]. Available: http://www.arcsight.com/solutions/solutions-cef/. [Accessed 13 Feb 2012].

[313] S. Ali and T. Heriyanto, BackTrack 4: Assuring Security by Penetration Testing, Packt Publishing, 2011.

[314] Trend Micro, Inc., "OSSEC v2.6.0 documentation," 2010. [Online]. Available: http://www.ossec.net/doc/log_samples/index.html. [Accessed 13 Feb 2014].

[315] J. Ritter, "ngrep - network grep," 2006. [Online]. Available: http://ngrep.sourceforge.net/download.html. [Accessed 13 Feb 2014].

[316] Department of Homeland Security, "Government Forum of Incident Response and Security Teams (GFIRST)," 2014. [Online]. Available: http://www.us-cert.gov/government-users/collaboration/gfirst. [Accessed 14 Feb 2014].

[317] The MITRE Corporation, "Threat-Based Defense: A New Cyber Defense Playbook," Jul 2012. [Online]. Available: http://www.mitre.org/publications/technical-papers/a-new-cyber-defense-playbook. [Accessed 16 Feb 2014].

[318] Wikimedia Foundation, Inc., "Kerberos (protocol)," 11 Feb 2014. [Online]. Available: http://en.wikipedia.org/wiki/Kerberos_%28protocol%29. [Accessed 16 Feb 2014].

[319] E. Fitzgerald, "The Trouble With Logoff Events," 8 May 2007. [Online]. Available: http://blogs.msdn.com/b/ericfitz/archive/2007/05/08/the-trouble-with-logoff-events.aspx. [Accessed 16 Feb 2014].

[320] R. F. Smith, "The Key Difference between "Account Logon" and "Logon/Logoff" Events in the Windows Security Log," 20 July 2011. [Online]. Available: http://www.eventtracker.com/newsletters/account-logon-and-logonlogoff/. [Accessed 16 Feb 2014].

Appendix A
External Interfaces

Regardless of where the SOC resides in relation to the constituency, it interacts with a variety of different entities on a regular basis. Some of these parties can be adjacent to the SOC physically and organizationally; others may be spread throughout the constituency, located down the hall, or on the other side of the world. As a reference to the reader, we provide a baseline set of definitions of who we're talking about and their respective function as it relates to the SOC. We cite these parties throughout the book.

In Table 25 we describe the parties a SOC has regular contact with, whether they are part of the constituency, and how those parties' roles support or relate to the SOC mission. Note that parties in bold typeface are those with whom a SOC often has the most frequent interaction.

Table 25. SOC Touch Points

Who	Inside or Outside of Constituency	Description and Function (as it relates to the SOC)
Constituency Chief Executive Officer (CEO)	Inside	Ultimately responsible for constituency mission, delegating key authorities to SOC, will express interest in some of the most severe incidents
Chief Information Officer (CIO)	Inside	Oversight over all IT, sometimes including IT security; will request SA and regular status updates from the SOC
Chief Information Security Officer (CISO)	Inside	Focused on full scope of cybersecurity; will want higher fidelity reporting and updates than the CIO; may wish (or actually have) control over what the SOC does
Information Systems Security Manager(s) (ISSM)	Inside	(In government organizations) Responsible for the day-to-day cybersecurity of a portion of the constituency; exerts some control over risk decisions about systems under their purview, particularly from an assessment and authorization perspective (if such a process is used)
Information Systems Security Officer (ISSO)	Inside	(In government organizations) Boots-on-the-ground presence across the constituency; responsible for working with users and sysadmins daily; can be instrumental in running incidents to ground and cleanup; plays role in audit review, which may create limited overlap with SOC mission. Some incidents may be handed over to ISSOs, such as routine computer misuse.

Table 25. SOC Touch Points

Who	Inside or Outside of Constituency	Description and Function (as it relates to the SOC)
IT Engineering	Inside	Large variety of staff that are responsible for design and development of systems and networks in the constituency; the SOC must stay on top of what they're deploying; may influence how networks and systems are instrumented to support intrusion detection
CND Engineering	Inside	IT engineering subgroup specifically responsible for acquiring, engineering, and deploying new SOC tools and upgrades; should be part of the SOC itself
Network Ops Center (NOC)	Inside	SOC's counterpart for network operations; can help find tip-offs for intrusions and deploy countermeasures; responsible for maintaining near 100 percent availability of networks and services, sometimes at odds with security
IT Help Desk	Inside	Who to call when something goes wrong with your computer; regular source of incident tip-offs
Users	Inside	Normally, call the help desk when they have an IT problem, but a well-advertised SOC can get direct calls when someone suspects they have a potential incident.
Business Unit Executive	Inside	Responsible for the full-scope mission or business area of large segments of the enterprise, they care when a system goes down or there is a breach of security.
System Owner	Inside	Responsible for a program or line of business containing many IT assets
System Administrator (sysadmin)	Inside	Performs hands-on operation and maintenance of IT assets; when there is an incident, usually the one who can help establish the facts of what happened and rebuild systems; also responsible for assessing the impact of countermeasures that the SOC recommends or directs
Business Unit Ops Center	Inside	Some business or mission areas will have an ops floor; the floor lead will usually have full-scope authority over all of the systems under their purview; if the SOC finds something that impacts the business unit's mission, the ops director will be one of the major points of contact for information flow, coordination, and response actions.
Counterintelligence (CI)	Inside/ Outside	Some constituencies have a unit specifically focused on preventing and finding threats against the people and mission of the constituency, such as espionage. Some of the most important incidents a SOC will uncover can be CI-related, requiring close coordination between CI and the SOC. CI usually has investigative authorities.

Table 25. SOC Touch Points

Who	Inside or Outside of Constituency	Description and Function (as it relates to the SOC)
Inspector General (IG)	Inside/ Outside	(In government organizations) Responsible for finding cases of waste, fraud, and theft, along with general auditing functions; some of the incidents a SOC finds will fall under their purview. IG usually has investigative authorities.
Legal Counsel	Inside	Provides legal advice to executives and members of the constituency. Some incidents found by the SOC will be referred to them. Also consulted to ensure a SOC's monitoring and data handling practices are legally sound, such as from a privacy perspective.
Law Enforcement (LE)	Outside	Federal, state, and local badge-wearing, armed crime fighters. Incidents found by a SOC may be referred to various LE authorities, if the situation warrants, but usually only after consultation with other legal counsel.
Auditors	Inside/ Outside	Third-party organizations responsible for reviewing various aspects of constituency finances and IT security controls. Auditors will regularly examine documentation and policies pertaining to SOC operations.
Blue Team	Inside	Performs full-knowledge assessments of constituency networks and systems, looking for security weaknesses. Sometimes staffed by people who normally work for the constituency organization. SOC should know about details of their assessment activity (such as network scans) in advance. Their results inform monitoring efforts of the SOC. May be based out of the SOC.
Red Team	Inside/ Outside	Performs no-knowledge simulations of an attack against constituency members with specific objectives in mind. Sometimes staffed by people who normally work for the constituency organization. Usually authorized by the chief executive without notification of other parties such as the SOC. Results should inform monitoring efforts of the SOC. Operations may be based out of the SOC.
National or Government-wide SOC	Outside (above)	Coordinating SOC for an entire nation, its entire government, or some large section of government (such as a branch or large department) whose constituency includes many other SOCs. Typically provides a range of services to member SOCs but may also have some operational authority over them. Their operational directives may consume significant resources of subordinate SOCs.

Table 25. SOC Touch Points

Who	Inside or Outside of Constituency	Description and Function (as it relates to the SOC)
Partner SOCs	Outside	SOCs with different constituencies, often in the same area of government, business, industry, education, or geographic region. Seen as a peer and "brother in arms," can be invaluable resource for heads-up on vulnerabilities, adversary TTPs, best practices, and tools. Leveraging these connections can help a SOC progress by leaps and bounds.

The SOC must coordinate its operations with many of these groups on a daily, weekly, or monthly basis, especially in response to incidents. Nurturing strong relationships helps a SOC execute its mission, especially when it may be lacking authorities or resources. On the other hand, many of these parties assert their own opinion when an incident occurs, which invariably presents as many challenges as it does opportunities. As a result, many SOCs find themselves in the middle of a political vortex—they must balance resources between coordinating with external parties and executing the CND mission.

Appendix B

Do We Need Our Own SOC?

Not every constituency needs its own incident response team. This need must be evaluated against a number of factors, including constituency size and IT budget. If a freestanding SOC isn't warranted, such as the case for small businesses [286], other options such as outsourcing may be an option. In this section, we evaluate factors that impact whether a constituency needs its own SOC, and discuss options for those that don't.

B.1 Assessing the Constituency

There is no industry-standard guideline for knowing whether to in-source incident detection and response. Therefore, we have come up with one as a starting point. In the following section, we have a worksheet (See Table 26.) that will help determine whether a given constituency can support a SOC or whether the constituency should pursue CND services from an external entity.

Consider the qualities of the constituency when filling out the worksheet. For questions 1–7, if the answer is "yes," give yourself one point; if not, zero points. At the second line from the bottom of the table, enter the number of thousands of hosts in your constituency. Multiply the number of thousands of hosts by the points subtotal, giving the total number of points at the very bottom.

As a general guideline—and this is where different experts on SOCs may have differing opinions—we pick a rough threshold of 15. Organizations scoring well above 15 are more likely to warrant a SOC. Those that score well under 15 may be better served by an ad hoc security team model or outsourced monitoring. An organization that scores right around 15 may look to other factors such as resourcing or organizational risk tolerance. Additionally, an organization's score is a *loose* indicator of the size and resources its SOC should have. In other words, an organization with a score of 200 probably needs a bigger SOC than an organization with a score of 20.

Let's consider two example organizations, score them, and examine the results. (See Tables 27 and 28.)

Big Toy Manufacturing, Inc. doesn't have a very large enterprise, its IT budget isn't very large, and it doesn't have a lot of risk factors that increased its score. Nonetheless, 2,000 hosts is not insignificant. With a score of 6, it doesn't fall into the range that strongly suggests a dedicated SOC. Either an ad hoc/decentralized SOC composed of members of its IT staff or outsourcing its CND needs to an MSSP might be appropriate. It probably won't

Table 26. Scoring the Need for a SOC

Item	What	Answer	Points
1	Give yourself 1 free point because you will have an incident at some point in time.		1
2	Has your constituency detected an incident that had a measurable impact on the mission or came at a significant cost within the last six months?		
3	Is there a perception that your constituency faces a targeted external cyber threat beyond the normal Internet-based opportunists such as script kiddies?		
4	Does your constituency serve a high-risk or high-value business or mission *and* is that mission heavily dependent on IT, such as finance, healthcare, energy production, or military?		
5	Does your constituency offer IT services to directly connected third parties in a B2B, B2G, or G2G fashion?		
6	Does your constituency serve sensitive or privacy-related data to untrusted third parties through some sort of public-facing portal such as a Web application?		
7	Does your constituency retain sensitive data provided or owned by a third party, such that the constituency faces significant liability if that data is stolen or lost?		
	Subtotal		
	How many thousands of hosts are in your constituency?		
	Multiply the subtotal by the number of thousands of hosts in your constituency. This is your total.		

need a large tool infrastructure, but a handful of a small log collection appliances and some host configuration monitoring tools might make sense.

Big Government Agency has a substantial enterprise with 20,000 hosts. While it doesn't directly deal with the public, it is the custodian of sensitive data that is owned by other entities. If there were breach of its systems, there's a good chance it would end up in the newspaper. As a result, with 80 points, it seems pretty clear it needs a SOC with a dedicated team of analysts and monitoring tools.

B.2 Outsourcing

If our score was less than 15 (or so), would the constituency warrant a standing SOC? In this case, we have a few options:

Table 27. Example #1: Big Toy Manufacturing, Inc.

Item	Explanation	Answer	Points
1	Give yourself 1 free point because you will have an incident at some point in time.		1
2	We just had to clean up a major botnet infection from hundreds of our Intranet hosts. It took us weeks to clean it up.	Yes	1
3	None that we're aware of.	No	0
4	No, all our production is done in Taiwan by third parties; we design toys, test them for safety, and handle sales.	No	0
5	No.	No	0
6	No.	No	0
7	We also design toys and toy parts for other companies. They give us their designs before they're launched, which are considered trade secrets.	Yes	1
Subtotal			
	We have roughly 2,000 IPs in our enterprise.		
	3 * 2 = 6.		6

1. If the constituency is part of a larger organization such as a government agency, business conglomerate, or large multicampus university, CND could be taken care of by the SOC for the parent organization. With large departments structured into subordinate bureaus, agencies, or offices, this often makes a lot of sense. This could be viewed as a form of outsourcing, but, in some cases, no money will change hands because of existing organizational and budgetary relationships.

2. Integrate security operations into an existing organization such as IT operations or a NOC (if one exists) or, perhaps, under the CIO. In this arrangement, we follow the security team model from Section 2.3.2, where there is no standing independent CND capability. Obviously, in this model, there is a significant risk that incidents will go unnoticed, or that if they are noted, they will not be dealt with in the most efficient, effective, or comprehensive manner.

3. Outsource CND to an MSSP. In this arrangement, the constituency pays a third party to monitor its enterprise, provide incident response, and, possibly, take on other services from Section 2.4, such as vulnerability assessments. Although we have a dedicated team of CND professionals, they are probably not tuned into the mission and internals of the constituency because they are neither part of, nor

Table 28. Example #2: Big Government Agency

Item	Explanation	Answer	Points
1	Give yourself 1 free point because you will have an incident at some point in time.		1
2	We're not aware of any major incidents as of late.	No	0
3	We're not sure, but no one has done a serious threat assessment of our mission as it relates to IT and cyber.	No	0
4	If our systems go down, our ability to process paperwork and keep our employees productive grinds to a halt .	Yes	1
5	Yes, we have several services that are directly tied into other government agencies.	Yes	1
6	No, we don't have a major Web presence other than a generic website that says who we are and what we do.	No	0
7	Yes, we share sensitive data about our citizens with other government agencies regularly through our databases.	Yes	1
Subtotal			4
	We have roughly 20,000 IPs in our enterprise	20,000	20
	4 * 20 = 80.		80

located near, the constituency. While their CND skill set and focus may be strong, their response time, their ability to influence policy, and insight into the mission may suffer compared to the ideal situation.

4. Hire a contractor who specializes in CND solutions to operate a SOC in-house. In this scenario, the SOC ops floor is located in the constituency's physical space but is staffed 100 percent by a third-party contractor. This arrangement has most of the pros and cons of the MSSP model, but it is somewhat muted compared to full-blown outsourcing (at potentially higher cost). Careful attention must be paid to defining a good contract, addressing issues such as cost model, personnel vetting, and technology investment decisions. In-house contracting tends to suffer from political friction between the SOC contractors and other stakeholders in IT ops, especially if they're on separate contracts, and because the SOC contractors often lack authorities to prevent or respond to incidents on their own initiative.

For more information on outsourcing to an MSSP, see [287].

Appendix C
How Do We Get Started?

If we have established the need for a SOC, the next logical question is, "How do we stand up a new SOC?" When we stand up a new capability, various priorities compete for our time and energy. The purpose of this section is to sort these priorities into different phases of SOC creation and growth, introducing many of the topics covered throughout the ten strategies. For more information on standing up a SOC, see [288] and Chapter 2 of [15].

Before we discuss the roadmap to standing up a SOC, here are some tips for success. Every SOC is different, so the timelines and order of priorities will differ; the following serves as a starting point and presents an ideal timeline for SOC stand-up:

- Ensure expectations and authorities of the SOC are well-defined and recognized from the start, especially from those in the SOC's management chain.
- Do a few things well rather than many things poorly; shun activities that can be easily or better performed by other organizations.
- Beg, borrow, or steal as much as possible:
  - Assimilate existing CND or CND-like capabilities into the SOC.
  - Leverage existing technologies, resources, and budget to help get started.
  - Don't let the initial influx of resources detract from the importance of a permanent budget line for people, capital improvements, and technology recap.
- Focus on technologies that match the threat and environment and act as a force multiplier; avoid getting caught up in "technology for its own sake"; extract the maximum amount of value from a modest set of tools.
- Having a flashy, well-organized ops floor isn't just for the analysts—it also keeps money flowing from IT executives. Having an advanced SOC is a point of pride for many seniors, and this starts with what they see when they walk onto the ops floor.
- Enable the rock-star analysts to lead all aspects of the SOC in a forward direction through continual improvements to processes and automation.
- Ensure strong quality control of what leaves the SOC from day one. Gaining trust and credibility is a big challenge, considering that rookie mistakes can easily undermine progress and stakeholder trust.
- Tune into the constituency mission, in terms of monitoring focus and response actions.
- Ensure each aspect of the SOC is given due attention. Start with a careful selection of the best people the SOC can attract, given budgetary constraints.

In Figure 32, we summarize the triad of CND that is of keen interest to new SOCs.

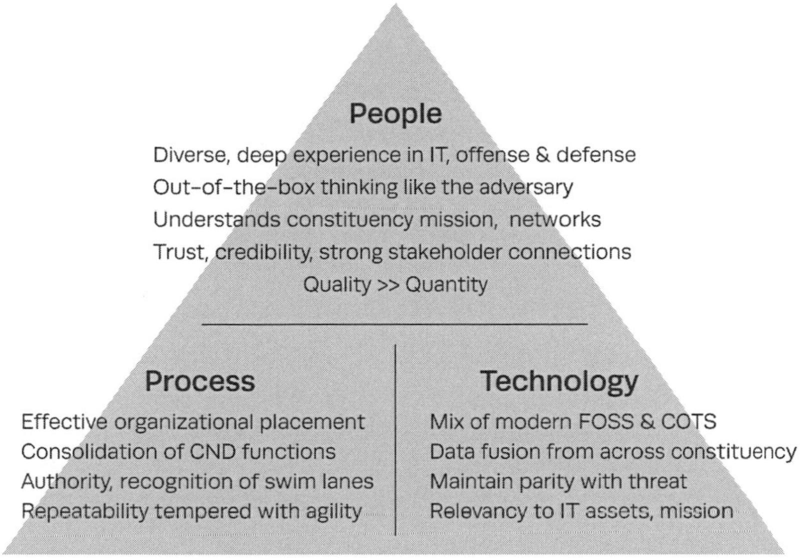

Figure 32. People, Process, Technology CND Enablers

C.1 Founding: 0 to 6 Months

In the beginning, there was no SOC—most likely, only pockets of CND being practiced across the constituency and a desire by seniors to "keep us out of the newspaper" or "defend the mission." From the time the decision is made to create a SOC, we have the following initial priorities:

- Form the team that will begin constructing the SOC, including its ops floor. Base this on existing experts in cybersecurity and CND, possibly along with the first hires to the SOC or outside consultants.
- Define the constituency.
- Ensure upper management support.
- Solicit input on which problems the SOC should solve and which capabilities are needed from constituency seniors.
- Write the SOC charter; get it signed by the constituency CEO or CIO.
- Collect CND best practices from literature and other SOCs.
- Secure funding for people and technology, based on a rough order of magnitude budget.
- Select a team organizational model (Section 4.1.1).
- Find the right place for the SOC in the constituency org chart (Section 5.1).

- If possible, begin the hiring process (Section 7.1 and Section 7.2), especially for lead analysts and engineers who can support the initial build-out of the SOC.
- Find a place for the SOC ops floor (Section 4.3) and begin its build-out.
- Find a place for the SOC systems (Section 4.3.2) that is physically near the ops floor or can be effectively managed remotely with little "touch" labor.
- Identify existing people and tools that could be brought into the SOC.

C.2 Build–Out: 6 to 12 Months

We hope to have some time between when a call is made for the creation of a SOC and when it is expected to assume sustained operations. During this time, our priorities are to formulate a detailed vision of full capability and focus on building toward that goal. After six months we should have a small team dedicated to standing up the SOC, allowing us to build a lot of our initial capabilities.

- Write and socialize the SOC CONOPS.
- Determine the initial team org chart (Section 4.2).
- Determine which capabilities to offer, at least initially (Section 6), working with constituents to identify areas where the SOC can provide the most added value.
- Make (or maintain) contact with other SOCs (either those that are operationally superior or those that are seen as peers) in similar areas of government, education, commerce, or geographic region.
- Build requirements for, evaluate, acquire, and pilot essential monitoring capabilities (Section 8.2, Section 8.3).
- Deploy a pilot monitoring package at a few enclaves or network perimeters physically close to SOC, thereby giving the first hires an initial monitoring capability to focus on.
- Build requirements for, evaluate, and deploy initial data aggregation and analytic capabilities such as an LM appliance or SIEM (Section 8.4).
- Begin hiring staff in large numbers, aiming for 50 percent capacity in the 6- to 12-month window.
- Perform the majority of the build-out of the SOC enclave (Section 10.2) and the ops floor.
- Integrate existing CND or CND-like technologies, processes, and personnel into the SOC, such as existing log collection or vulnerability scanning.

C.3 Initial Operating Capability: 12–18 Months

If ops floor construction and tool acquisition have proceeded according to plan, we should now have at least a part of the physical space ready for operations. In addition, members of the SOC team should now be showing up for duty. We can now legitimately begin operations, at least in an ad hoc manner. Then:

- Finish hiring staff in large numbers, aiming for 90 percent capacity in the 12- to 18-month window.
- Leverage newly acquired tools and ops floor space to begin creating a monitoring and analysis framework, ensuring that key information and tools are at the analysts' fingertips (Appendix F).
- Begin development of SOC SOPs and notional ops tempo (Appendix D).
- Begin development and socialization of lower level authorities (Section 5.1).
- Begin regular analyst consumption and fusion of cyber intel into monitoring systems (Section 11.7), supporting an initial SA capability.
- Identify and recruit TAs at remote sites, if appropriate (Section 4.3.4).
- Deploy production sensor capabilities to the initial set of monitoring points (Section 9.1).
- If capability doesn't already exist, begin gathering log data; if it does, ensure feeds critical to CND are part of the mix (Section 9.2).
- Begin advertising the SOC, including establishment of a SOC Web presence.
- Establish an incident tracking/case management capability (Section 12.2).
- Begin sustained detection, analysis, and response operations (Section 12.1).

C.4 Full Operating Capability: 18 Months and More

Each SOC has its own definition of full operating capacity (FOC), but generally speaking, at this stage, we should see the SOC rise to the full scope of the mission. Whereas, in the beginning of operations, many tasks were performed in an ad hoc manner, we should now transition the SOC into a sustainable ops tempo, consistent with a growing set of SOPs.

- If necessary and not already established, expand working hours (possibly to 24x7 operations) (Appendix D).
- Establish practices to maximize quality staff retention and growth (Section 7.2).
- Demonstrate the value added to constituency mission by SOC's handling of cyber incidents.
- Solicit feedback from constituents.
- Adjust operations procedures and capabilities as necessary, given the deltas between the initial vision of the SOC and the operational, resourcing, and policy realities.
- Deploy monitoring capabilities to an expanded set of monitoring points as appropriate.
- Expand log collection and analytics.
- Build up data filtering, correlation, triage, and analysis automation techniques.

- Expand SOC influence into areas of policy, user awareness, and training, if appropriate.
- Establish regular sharing of cyber intel and tippers with partner SOCs and SA with constituents.
- Consider measuring SOC effectiveness against a holistic metrics program (See [289].) and annual or semiannual Red Team exercises.

Even though a SOC has achieved its FOC, it will almost certainly take a bit longer for it to become fully "mature," since its mission and ops tempo are always changing. In Appendix G we cover the characteristics of a healthy, mature SOC.

Appendix D
Should the SOC Go 24x7?

The adversary works 24x7. Should the SOC as well? In order to answer this question, we must carefully examine the scope of the SOC's mission, its staffing resources, and the daily/weekly patterns of activity across the constituency.

For many SOCs, going 24x7 isn't an all-or-nothing decision. Many SOCs that keep a 24x7 watch, staff only Tier 1 around the clock—most other functions are performed during regular business hours and on an on-call basis. Some SOCs maintain a 12x5 watch with eight-hour skeleton staffing on weekends. Some of the largest SOCs staff all branches 24x7, with the bulk of the resources present during regular business hours (e.g., with asymmetric staffing). Finally, some SOCs that have multiple ops floors will stagger shifts between them—as in follow the sun or backup/contingency ops floors in disparate time zones, although this is less common.

Finding the right staffing plan can be a challenge; the plan depends on a number of considerations, including:

- What is the size of the constituency and what are its normal business hours? Does its user population have after-hours access to IT resources?
- Does the constituency's mission encompass 24x7 operations that depend on IT, which, if interrupted, would significantly imperil revenue or life?
- How big is the SOC's staff pool? Are there only four people, or is it funded to support a team of 50 or more analysts?
- Is there a specific set of targeted or advanced threats against the constituency that suggests intrusion activity is likely to happen outside of normal business hours?
- Have members of the SOC come in after hours to handle an incident, or have they discovered an incident outside SOC duty hours that could have been prevented if someone were there to catch it? Has this happened several times?
- If an attack occurred during off hours and there was an analyst there to notice it, are there resources outside the SOC that could stage a meaningful response before the following business day?
- Is the host facility open 24x7? If not, can the SOC be granted an exception?

There is a potential stigma associated with not keeping the lights on all the time. Some SOCs are considered a legitimate ops center only if they function 24x7. Moreover, a SOC that isn't 24x7 must catch up every morning on the previous night's raft of event data; for those that don't staff on the weekends, this is especially challenging on Monday.

Given that, let's take a look at some of the realities that come with around-the-clock staffing:

- 24x7 SOCs must maintain a minimum staff of two analysts at all times:
  - Two-person integrity is a best practice in monitoring since having only one person there with access to a lot of sensitive data and systems can present problems, no matter how well-vetted the employees.
  - There are logistical and safety concerns with keeping the floor staffed and secured when someone needs to leave the room.
  - With multiple analysts always on shift, they can cross-check each other's work.
  - Being the sole person on shift can be very lonely and monotonous.
- Each 24x7 seat requires roughly five FTEs, including fill-in for vacation and sick leave.
  - This is very expensive compared to 8x5, 12x5, or even 12x7 staffing.
  - Assuming a minimum of two filled analyst seats, that means roughly 10 FTEs.
  - Therefore, taking a SOC from 8x5 to 24x7 requires an increase of at least eight FTEs just for Tier 1.
- Despite two-person minimum staffing, it's easy for unattended analysts on nights to spend more time watching TV and browsing social media sites than analyzing.
  - This is a common occurrence, especially when the ops floor is physically isolated.
  - It is important to have regular deliverables/work output from those on night shift and regular feedback regarding what happened during the day.
  - Because night-shift staff have far fewer interruptions, it may sometimes be effective to give them unique tasks that require several hours of focused work, such as in-depth trending or cyber intel analysis.
  - Collocating the SOC ops floor with a NOC ops floor may help, especially with loneliness and management supervision.
- Because analysts on night shift almost universally feel underinformed and unrecognized, the feeling of "out of sight, out of mind" can crop up.
  - Casual information sharing occurs less when only a few positions are staffed.
  - The night shift does not see the fruits of their labor as much because detailed analysis, response, and feedback usually occur during normal business hours.
  - These issues can be largely offset by rotating staff between days and nights every three to four months and by keeping a lead analyst on staff during the night.

As we alluded to above, there are multiple options for expanded staffing that don't incur the full cost of going 24x7:

- Staff only certain portions of the SOC 24x7, such as Tier 1; leave other sections with a designated "on-call" pager or cell phone.

- 12x5. Expand operations beyond 8x5 so that there are SOC analysts on shift during the bulk of time that constituents are logged in, such as 6 a.m. to 6 p.m., assuming that the constituency resides primarily in one or just a few adjacent time zones.
- 12x5 plus 8x2. Add one shift (8 or 12 hours) of two or three analysts during weekends. This eliminates the problem of clearing a weekend's worth of priority alerts on Monday and provides coverage if the constituency performs business operations on weekends.
- Outsource. If the SOC follows a tiered organizational model, it could, during off hours, hand off operations to the parent coordinating SOC or sister SOC—assuming they are 24x7, can easily access the SOC's monitoring systems or data feeds, and are able to familiarize themselves with the SOC's constituency mission and networks.

During each shift, watch-floor analysts may be encouraged to keep track of important pieces of information in a **master station log** (e.g., phone calls, interesting events, visitors to the SOC, or anything else out of the ordinary). This log can be instrumental in reconstructing timelines of events, enforcing accountability, and demonstrating SOC due diligence.

At the end of each shift, the leaving team will perform what is often referred to as shift **passdown** or shift **handoff**, whereby the outgoing team briefs the incoming team on various issues of note seen during their shift. It is a good practice to use both the master station log and a **passdown log** to formally document this information handoff. Again, the purpose of this process is to enforce continuity of ops, support nonrepudiation and accountability of the floor staff's actions, and serve as a historical record.

In non-24-hour environments, this passdown log should probably still be maintained, although any sort of person-to-person handoff will need to work differently due to non-continuous staffing. Regardless of the staffing model, here's what can go in the passdown log:

- Names of SOC operations staff on duty
- Issues passed down from the previous shift, especially those mentioned verbally and not captured in the previous passdown log
- Case IDs opened and closed during the shift
- Tips and referrals from other parties such as the help desk or users
- Any issues escalated to parties outside the SOC
- Sensor or equipment outages seen
- Any sort of anomalous activity seen, especially if it has not yet spawned a case
- Any anomalous activity that was seen that requires the incoming shift to continue analysis or escalation procedures
- A check-off of duties that the team was required to accomplish

- Details on any tasks assigned to the shift that were not completed or need further attention
- Anything else of significance that was encountered during shift that isn't covered in the SOC's SOPs.

It is best to have a standard passdown log template that each shift uses, which is usually filled out by the shift lead or team lead on duty. While the log may be captured electronically, it is important to print the log and have all analysts on the outgoing team and the incoming shift lead sign it before the outgoing team leaves for the day. This is key to maintaining accountability for what was done and ensure that nothing is dropped.

Appendix E
What Documents Should the SOC Maintain?

The SOC lives in the middle of a political vortex; meanwhile, it must maintain consistency in operations and cope with high turnover. One of the best ways of dealing with these realities is to maintain documentation describing various aspects of the SOC's mission and operations. This is especially handy when additional scrutiny is focused on the SOC or when new employees must be trained. Table 29 lists each of these documents—what they are, what they say, and why the SOC needs them. SOC leadership should evaluate its own mission needs against this potential document library and consider how often it needs to revise each—some every two to three years; others, maybe monthly.

Table 29. Document Library

What	What It Says	Why You Need It
Charter Authority	The scope of the SOC's mission and responsibilities and the responsibilities of other groups with respect to CND, signed by the chief executive of the constituency	Always keep this handy. While most groups cooperate willingly, sometimes the SOC must wield its authorities in order for others to support incident response or prevention.
Additional Authorities	Detailed authorities and clarification about SOC mission and touch points that fall outside the charter. It fills in certain details that the charter leaves out (e.g., what the SOC can do in response to or prevention of an incident) or describes additional capabilities taken on after the charter was signed. Can be signed by someone in the SOC's management chain.	Same reason as the charter—these will clarify what the SOC can and should do and what other orgs are obligated to do in helping the SOC. Good examples include incident escalation and swim lanes. See Section 5.1.
Mission and Vision	Two crisp statements/slogans saying what the SOC does and what it is aiming for in the future	Helps orient members of the SOC toward a common set of objectives and, in just a few sentences, helps external parties understand what the SOC does

Table 29. Document Library

What	What It Says	Why You Need It
CONOPS	Covers not only the what and why of the SOC mission, but also the how and who. This includes the roles and responsibilities of each of the SOC's sections, the technologies it uses, and its ops tempo, inputs, and outputs. While it may articulate escalation flowcharts for major incidents, it does not get down to minute details of specific checklists.	Essentially the one-stop show for members of the SOC to understand how the SOC functions, without necessarily covering incident or job specifics. Some SOCs choose to split this document into two pieces: one part for internal consumption and another for reference by other parties.
Shift Schedule and On-Call Roster	The shift schedule for the SOC, at least two weeks into the future, including who will be on each shift position and who from each section is the designated "on call" person for times of the day or week that that section is not manned	So staff knows who their relief will be and whom to call if they have questions about what happened on the previous shift
Incoming Incident Reporting Form	Constituents fill this out when reporting an incident to the SOC. It captures all incident details the submitter is able to capture—who/what/when/where—what systems were involved, symptoms were observed, time/date, and whom to call for follow- up. This form should be available on the SOC website. See Appendix E of [3] for examples.	Provides a consistent means for the constituency (users, help desk, sysadmins, ISSOs, etc.) to report potential incidents to the SOC
Incoming Tip-Handling SOP	Instructions for handling incoming incident tips: what data to capture, what to do next, whom to call, thresholds for further escalation, and the like	Ensures the right information is captured and correctly escalated. Tier 1 should closely follow this SOP every day; it ensures Tier 2 can respond effectively.
Escalation SOP	Sets thresholds and escalation paths for whom Tier 2 passes incidents to (security, LE, CIO, etc.). May be released to the constituency so everyone can understand who the SOC calls and under what circumstances.	Members of the constituency are very sensitive to who gets to know about which incidents and when. This ensures everyone knows who gets which incidents at what threshold.
Shift Passdown Form	Defines what information must be captured by the incoming and outgoing shift. In non-24x7 shops, this is often used, even though there is no "handoff" from one day to the next.	Ensures nothing gets dropped and major events are recorded; enforces accountability. See Appendix D for more details.

Table 29. Document Library

What	What It Says	Why You Need It
Artifact-Handling Process	Defines the process and steps SOC members must follow in accepting, collecting, storing, handling, and releasing digital and physical artifacts. May reference other legal guidelines for evidence handling. This document should probably be reviewed by legal counsel before approval.	Ensures that the SOC's hard work stands up in court, in the event an incident leads to legal action.
Monitoring Architecture	Articulates the details on where (logically or physically) the SOC's monitoring capabilities are located and how that data (PCAP, events, logs, etc.) is collected and stored. Should depict a detailed path from the end network all the way to the analyst. Some SOCs break this into two pieces: (1) a generic depiction of how the constituency is instrumented, and (2) a detailed diagram showing exact sensor locations; the former can be shared, the later should not.	Helps SOC members understand how their network is instrumented. Being clear on exactly where a sensor is tapped is critical because there are always subtle blind spots due to DMZ and routing complexities. It also helps SOC sensor and sysadmins troubleshoot downed feeds when they occur.
Network Diagrams (See Section 8.1.2.)	Depict the detailed network architecture of the constituency, usually showing user networks and server farms as clouds connected by firewalls and routers. Typically broken down into a series of large Visio diagrams that can be printed on a large-format printer. Regardless of whether a SOC maintains these diagrams, it should consider overlaying its sensor placement for internal tracking purposes.	Help members of the SOC understand the size and shape of the constituency, how data gets from point A to point B, where external connections are, and the connection between subnets and mission/business functions.
Internal CM Process	Defines how changes are made to SOC systems and documents (e.g., hardware and software installs/upgrades, IDS and SIEM signature changes, and SOP updates).	Ensures that rigor and consistency are enforced—with notification and visibility across the SOC for changes—while balancing agility in ops. For instance, a SOC should be able to turn a piece of cyber intel into a signature push in a matter of hours, but not without documents that notify analysts of the change.

Table 29. Document Library

What	What It Says	Why You Need It
Systems and Sensors Maintenance and Build Instructions	A series of documents that discuss how to maintain all key SOC systems and how to rebuild them in the event of corruption or hardware failure	While vendor manuals always help, a SOC will have many customizations, especially for homegrown solutions. For example, joining a system to a SAN requires work with at least three different products. It is easiest to distill this into a few pages of instructions rather than pointing sysadmins at 1,000 or more pages of product manuals.
Confidentiality Agreement/ Code of Conduct	Concise statement of the rules that define the expected behavior and prohibited activities of SOC staff, above and beyond other agreements they signed as part of the constituency. It will usually articulate the need for SOC staff to maintain strict confidentiality about case and privacy data and to avoid snooping outside the scope of legitimate monitoring duties. This document should be reviewed by legal counsel before approval.	It has been said that with great power comes great responsibility. Should a SOC team member do something seriously wrong, this document supports corrective and legal actions against that employee. It also demonstrates to external groups that the SOC takes its job very seriously and holds its people to a high standard.
Training Materials, Technical Qualification Tests, and Process	Articulates staff in-processing, necessary training, periodic recertification, and qualification tests. Leverages many of the documents in this table.	Serves two key functions: (1) orients new staff on the SOC mission, structure, CONOPS, and SOPs and (2) ensures that each team member is proficient with SOC tools.
Operational, Functional, and System Requirements	Detailed listing of all the needs the SOC has for its tools. Contains everything from sensor fleet management specs to capabilities of malware analysis tools. Can articulate needs at three levels: (1) operational (what business needs to be done), (2) functional (what features are needed), and (3) system (what are the specifics of the implementation).	Helps support intelligent acquisition for all SOC capabilities. Gives the engineers a concrete set of needs that must be satisfied. Helps the SOC ensure it's getting what it needs, especially if the engineers are not part of the SOC.

Table 29. Document Library

What	What It Says	Why You Need It
Budget	Allocates money for SOC staffing, software/hardware licensing, refresh and maintenance, expansion, and capital improvements for the SOC, during the current fiscal year and at least three years into the future. Recognizes different categories of money and considers both inflation and expec ted changes in SOC capabilities.	A SOC must plan and budget for its capabilities just like any other organization in government, industry, or education. Having this at hand (along with a crisp list of successes) will help the SOC defend its budget against constant scrutiny and potential cuts.
Unfunded Requirements	Succinct one- or two-page description of each capability not currently built into the budget. Will include what the SOC wants, what benefit the capability will provide, how much it will cost, and what will happen if the SOC doesn't get it.	Having these at hand will help the SOC claim money when it becomes available from other organizations. Many SOCs have to beg, borrow, or steal. Being the first one to respond with well-written unfunded requirements usually means beating out other organizations for a portion of the funds.
IDS and SIEM Signature/ Content List(s)	List of all the signatures and content deployed to each SOC sensor or analytic system (IDSes, SIEM, etc.). This should be contained within the tools themselves, which is usually easier and more efficient than a separate document. Custom signatures and analytics are especially important to document—what they look for and what analysts should do when their alerts pop up.	Helps analysts know what they're looking at and what to do with each fired event. This list should be scrubbed by sensor managers and other key SOC stakeholders on a regular basis, perhaps quarterly.
System Inventory	System host name, IP, MAC address, hardware type, location, and serial/barcode of all SOC assets, along with other hardware and peripherals	The SOC must be able to keep track of what it owns so nothing gets lost; inventory must be refreshed on schedule.
Short Mission Presentation	15- to 30-minute slide presentation about the SOC—its mission, structure, how it executes the mission, and key successes	Used to describe the SOC to non-technical audiences in conferences or for quick demos for visiting VIPs. Helps gain trust among stakeholders and partners.
Long Mission Presentation	Longer (one to two hours) technical presentation highlighting SOC successes and TTPs, monitoring architecture, and future initiatives	Used for technical audiences such as other SOCs. Key for making connections, gaining street cred.

It is usually not enough to have these materials lying around in soft copy or hard copy. Most medium- to large-sized SOCs devote staff resources to knowledge management (e.g., keeping track of all these documents, including updates, in an orderly manner). The most popular means do to this is to post various information to a Windows file share in the SOC enclave, though many SOCs will augment this with Microsoft SharePoint or a wiki.

Appendix F

What Should an Analyst Have Within Reach?

Members of the SOC must draw on an ever-expanding universe of resources and information in their day-to-day job. Making sure these resources can be called upon at a moment's notice keeps the SOC running efficiently, saving critical seconds and minutes when tracking down an incident. All of these resources should be located either logically or physically close to the analysts—on their desks, on their workstations, or somewhere near them—in an organized and neat fashion. In Table 30, we list what resources are needed, how important (generally) they are, and why the operator needs them.

Table 30. Analyst Resources

What	Importance	Why
General Internet access	High	Access to CND websites, news, public–facing email, general troubleshooting, and external collaboration
Unattributed/unfiltered Internet access[1]	Medium	Was that user really surfing porn? Where did this piece of malware come from? These questions need to be answered daily, without placing the constituency at risk or tipping off the adversary.
Access to constituency main business network with email, office automation software	High	An analyst requires regular communication with constituents and the ability to conduct general business.
Employee directory(ies) for entire constituency	High	An analyst requires regular communication with others in the constituency.
Access to user–submitted incident reports; read and (possibly) write access to the SOC externally facing website/portal	High	A central point of communication between the SOC and the constituency, this is often where constituency users go to submit potential incident information.
Access to SOC network from a robust workstation that is not used to connect to the main constituency network or the Internet	High	As discussed in Section 10, the SOC's monitoring infrastructure should be placed in a well–protected enclave. Therefore, the analyst should access the bulk of SOC monitoring tools from high–performance workstations on the SOC network.

1 A number of details must be considered when deploying and supporting a truly unattributed Internet connection that are beyond the scope of this book. An unattributed Internet connection assumes that it cannot be traced back to the constituency, and there are no content filtering technologies on it, unlike the constituency's main Internet gateway that one expects has a robust content filtering solution in place.

Table 30. Analyst Resources

What	Importance	Why
Multilevel desktop consolidation system or KVM switch, if more than two or three different desktop systems are needed	Low	Most SOCs can get the job done with two workstations for each analyst: one for the SOC network and the other for Internet and constituency network access. Other SOCs will need more than this, in which case a KVM switch is necessary to reduce desktop clutter for some analysts.
General user privileges to the SIEM and/or log aggregation consoles	High	A SIEM usually serves as the hub for alert triage and event analysis; an analyst should be able to spend more time with the SIEM console than any other tool or system.
General user privileges to IDS consoles or other out-of-band monitoring systems	High	Some IDS consoles may have details beyond what SIEM can (or should) capture; in this case, analysts will also need access to these.
General user privileges to all in-band monitoring device consoles, (e.g., HIPS and NIPS)	Low	Same reason as out-of-band IDS consoles: they may offer more detail or options than what is collected by the SIEM.
Complete and current signature list for all IDSes, with signature descriptions and signature syntax	Medium	The analysts should know the policy that is currently deployed to all monitoring equipment, including the description of each alert and the exact syntax of the signature, if available.
Complete content list with descriptions for all production SIEM content	Low	Same rationale as IDS signatures but less important due to the comparatively small number of SIEM content/use cases and their comparative complexity
For every IDS alert seen, the raw event details, the signature (or signature description) that triggered it, the raw PCAP for that event, and supporting NetFlow	High	Without this data, IDS alerts are meaningless. The SIEM or IDS console should contain or link directly to all of these items. (See Section 8.2.)
Access to historical alert and PCAP data for Tiers 1 and 2 within the time frames defined in Table 18, Section 9.2	High	Different parts of the SOC will need access to different slices of data the SOC collects for historical analysis, both to look for new incidents and to establish the facts about existing incidents. These needs are outlined in the referenced table. This should include both the PCAP data itself and the tools necessary to view it.

Table 30. Analyst Resources

What	Importance	Why
Vulnerability scan results and/or on-demand vulnerability scanning tool	Low	Did that IDS alert hit a system that was actually vulnerable? What OS and services is this system running? Analysts will ask these questions regularly, and vulnerability scanners have the answers. Having access to both historical regular scan results and an on-demand scanning capability is best, but either one will help.
Network maps depicting major subnets and interface/peering points for the constituency (e.g., firewalls and IDS monitoring locations)	Medium	Analysts must understand the network they are monitoring. It helps to have a few key network maps (such as Internet gateways) posted on the wall of the ops floor. If the SOC is not responsible for network mapping, it's best to have read-only access to where these maps are stored on the constituency network.
Read-only access to asset-tracking database	Low	Complements vulnerability scanning data, especially when scan results are stale or unavailable. May also capture information about system owner, contact info, or supported mission.
Current firewall rule sets for all production firewalls in constituency	Low	Did that attack make it through the firewall? Should we be concerned that a given network is wide open? Current firewall rule sets answer these questions. Having read-only access to firewall rule sets helps, rather than having to ask for a download from the firewall managers.
Current router and switch configurations for all production network equipment in constituency	Low	In the same vein as firewall rules, analysts will often have to run down exactly where a system is located, and how it connects to other systems in (or outside) the constituency. In addition, router ACLs may be used to complement firewalls for simplistic security separation.
Incident tracking/ticketing database limited to SOC only	High	Analysts must be able to record pertinent details about incidents, attach events or other digital artifacts, and escalate that information daily to other members of the SOC.
Analyst collaboration forum/SharePoint/wiki and unstructured file share limited to SOC only	High	Members of the SOC will need to capture lots of unstructured or semistructured information not directly related to a given incident. Having both organized (wiki, SharePoint) and unorganized file share means of pooling these resources is important. In addition, a real-time chat room may be helpful. All of these resources should be on the SOC enclave.

Table 30. Analyst Resources

What	Importance	Why
Write access to current shift pass-down log (and/or master station log) and read access to past pass-down logs	Medium	Analysts should record their actions and events of note during their shift and summarize them at the end. That way, analysts on later shifts can review what happened and understand what issues require follow-up. Also helps support accountability.
Access to collaboration forums shared with sister SOCs	Medium	Messaging boards and collaboration forums where multiple SOCs share cyber intel, tippers, incident reports, and general cyber news are immensely helpful. These forums should be well protected and participated in by invitation only.
Standard public/commercial telephone	High	Same reasons anyone in business and IT needs a phone. Analysts will spend significant time on the phone every day; sometimes headsets are beneficial.
Secure telephone (e.g., secure telephone unit [STU]/secure terminal equipment [STE]), where applicable	Varies	If appropriate, the SOC may need a secure communications channel to people in the constituency or to other external organizations. In many cases, the sensitive nature of CND makes this even more important.
Contact information for all parties that are part of the SOC's incident escalation chain	High	Analyst will need to call TAs, sysadmins, ISSOs, and other parties that are both inputs and outputs to the SOC's incident escalation flowchart. These calls are often time sensitive, so, having up-to-date contact information readily available is key.
Personal (home and cell) contact information for all members of the SOC	High	Members of the SOC will get called after hours on a regular basis. Systems break, and analysts call in sick. Having this information printed on a laminated card that clips to a SOC member's physical security ID badge is a good idea.
Documents listed in Appendix E, except budget and requirements	Medium	Analysts will need to refer to SOPs, CONOPS, and other materials from time to time. While some of these documents are already listed in this table, having all the operationally relevant ones at hand is a good idea.
Secure document and media storage (where applicable)	Varies	The SOC will have to store sensitive data, forensic images, passwords to SOC systems, case data, and other materials in an appropriate manager. Chances are, at least one safe will be needed.

Table 30. Analyst Resources

What	Importance	Why
Emergency "go-bag"	Medium	Includes everything the SOC will need during an evacuation or catastrophic event. This includes contact information for all SOC members, rally location, shift schedule, flashlight, satellite phone, COOP activation playbook, and so forth. The floor lead should grab it on the way out the door during an emergency or fire drill.
Real-time network availability status dashboard	Low	It helps to get a feed of planned and unplanned network and system outage events across the constituency, provided by the NOC or IT ops.
Real-time cable news feed	Medium	No 24x7 ops floor is complete without one or two flat screens permanently tuned to a 24-hour news channel such as CNN or MSNBC. They help provide SA about events that may impact the constituency.
Vendor technical documentation on SOC technologies	Medium	The SOC should have references (either in hard copy or soft copy) for all SOC monitoring and analytics systems that analysts use.

In addition to all those listed above, there are resources that other members of the SOC responsible for Tier 2, Tier 3 (if it exists), cyber intel analysis, and trending will need access to, depending on their exact role. They are as shown in Table 31.

Table 31. Tier 2 Tools

What	Importance	Why
PCAP capture and manipulation and TCP/IP protocol analysis tools (WireShark, mergecap, etc.)	High	When evaluating an IDS event, an analyst can only establish the ground truth through full-session capture of network traffic. Along with the raw PCAP, an analyst needs tools to slice, dice, and analyze it.
Linux/UNIX system with Perl and open source tools (for text log processing) and plenty of scratch storage	High	While there are plenty of commercial tools to generate, transmit, store, parse, and analyze log data, there's nothing like having a Linux/UNIX shell prompt and a powerful scripting language such as Perl or text-parsing tools such as grep, sed, and awk.
Decompilation and static malware analysis capability (IDA Pro)	Medium	When examining machine(s) suspected of a compromise, there will likely be files that are suspect, but don't trigger traditional AV tools. As a result, an analyst will need to further examine them.

Table 31. Tier 2 Tools

What	Importance	Why
Malware detonation chamber and runtime analysis	High	An alternative approach to static analysis is runtime analysis—actually executing suspected malware and examining its behavior. This approach can also be automated to a limited extent (e.g., scripted malware "detonation chambers"). (See Section 8.2.7.)
Read-only hard drive imager	High	Even if a SOC doesn't do in-depth forensics or malware analysis, the most cursory inspection of media involved in an incident requires making a copy of the original. This hardware device (along with media image analysis tools) allows copying a hard drive without performing any write operations against it.
Media image analysis (FTK, Encase)	High	Once a hard drive image or other piece of media is acquired, these tools will support inspection and analysis of its contents.
Blue and Red Team out-briefs and detailed results	Low	For the same reason network maps and vulnerability scan results are desired, these provide a more holistic view of key parts of the constituency but are usually point-in-time and, therefore, must be augmented with updated network maps and vulnerability scan data.
All supporting authority documents listed in Section 5.1	Low	Knowing what the SOC has written authority to do or not do is important. SOC leads will refer to these on a semiregular basis. It's best to have them consolidated in one place.

Appendix G

Characteristics of a World–Class SOC

How do we know whether the SOC is doing well? Every organization is different, and there is no universal set of measures for SOC effectiveness. In this section, we describe the qualities of a SOC that has reached an ideal state of maturity, given the needs and constraints of its constituency. We draw on material presented throughout the book in articulating a target state for modern SOCs. While these qualities describe SOCs of any size or organizational model, we once again aim for common SOCs that:

- Serve a constituency of some 5,000 to 5,000,000 users or IPs
- Are members of their constituency
- Have at least shared reactive authority
- Has direct visibility into a large portion of the constituency
- Follow an organizational model that includes both centralized and distributed elements.

SOCs that fall outside this description (e.g., national-level and coordinating SOCs) will certainly be able to leverage elements of this section but may find that certain qualities won't apply. For instance, a mature SOC serving a small constituency may not be able to support advanced engagements with the adversary. Or, a mature SOC serving a very large constituency may not directly monitor constituency systems.

In describing a healthy, mature SOC, we start with the most basic elements of the SOC mission: prevent, monitor, detect, respond, and report—along with general programmatics, external connections, and training/career. SOC managers may certainly use these qualities as a basis to measure their SOC capabilities; however, we don't always go into detail on how to measure them. This is done best on a case-by-case basis.

One important caveat to recognize is that we are describing the ideal state of a world-class SOC. This state will never be reached in all regards, which is to say no one organization will ever receive an equivalent "100 percent" score. That said, we lay out these qualities as a target that a well-resourced SOC can shoot for.

Before we go any further, it is critical to recognize that the best way to measure overall SOC effectiveness is by running realistic drills and exercises against the SOC. Some exercises can include "tabletop" scenarios with the SOC and its partners, or exercising a COOP capability, if it exists. These are relatively low risk and can be done on a regular basis, perhaps annually. These, while useful, don't hit the nail on the head.

> **The best way to test a SOC is to measure the SOC's performance in response to an actual Red Team penetration of constituency assets.**

There really is no substitute for simulating a no-warning, full-fledged intrusion. The Red Team should actually execute an attack covering the entire cyber attack life cycle against a segment of the enterprise that is important to the constituency mission. Of course, this must be tempered with executing the operation so that it does not imperil actual constituency business or mission operations. The Red Team scenario should reflect the resources and motives of a true constituency adversary, whether an internal rogue actor or an external group.

For instance, a mock phishing attack against non-privileged users may exercise many elements of the SOC. This kind of exercise can be executed with greater frequency because it's relatively easy to do, has very low risk, and is much cheaper than a true Red Team exercise. But it does not fully exercise the SOC. What may be more useful is actually running a Red Team against a COOP or preproduction instance of a critical mission system. This should include not only reconnaissance and attack but persistence, privilege escalation, remote access, and, perhaps, data exfiltration. Incorporating inputs from SOC "TAs" in formulating such an exercise will enhance its realism and usefulness.

Whatever method used to assess the SOC, we hope that the SOC demonstrates qualities as described in the following sections.

G.1 Program

This section describes the qualities of the SOC's overall program that span multiple functional areas.

- The SOC is granted unambiguous authority to execute the CND mission for its constituency; for SOCs that are members of their constituency, this means that:
  - The document(s) that state their fundamental, high-level authority, such as a charter, carries the weight of the chief constituency executive.
  - The SOC has one or more additional documents that codify its more detailed authorities (those listed in Section 5.1), and also cover the roles and responsibilities of the SOC and its partner organizations with respect to the incident life cycle.
  - Any of these authorities or documents have been vetted by SOC management and follow the constituency's governance process.
- The SOC has direct control over a dedicated budget supporting its annual operations for both staffing, new technology capital investments, and technology refresh.

- The SOC incorporates the elements of CND described in <u>Section 3</u>, all under the SOC organizational structure:
  - Real-time monitoring and triage (Tier 1)
  - Incident analysis, coordination, and response (Tier 2 and above)
  - Cyber intel collection and analysis
  - Sensor tuning and management, scripting and automation, and infrastructure O&M
  - SOC tool engineering and deployment.
- The SOC has defined the set of capabilities it performs in support of its mission. (See <u>Section 6</u>.)
- The SOC has defined an internal organizational and management structure with clear division of roles and responsibilities. (See <u>Section 4.1</u>.)
- The SOC has balanced the need to maintain visibility out to the end asset and mission with the need to centrally consolidate knowledge and operations through a balance of centralized and distributed resourcing. (See <u>Section 4.1</u>.)
- The primary drivers or the SOC's day-to-day tasking, resource investments, and budget are its own internally defined priorities and SA.
- The SOC's day-to-day tasking, resource investments, and budget are driven only secondarily by external compliance.
- The SOC leverages a lightweight internal metrics program that:
  - Measures key aspects of SOC operational health, as determined by SOC management
  - Is used by SOC management to drive improvement to SOC process and performance
  - Is not widely recognized by SOC staff as being "gamed" or manipulated for purposes of false or inaccurate reporting
  - Consumes no more than 5 percent of the SOC's total labor through its implementation.
- The SOC is able to execute its mission within the time frames listed in <u>Section 2.8</u>.
- The SOC's mission is carried out primarily by personnel consolidated into a SOC ops floor (or, in COOP or tiered arrangements, multiple SOC ops floors).

G.2 Instrumentation

This section describes the qualities of the systems and procedures used to instrument the constituency for monitoring each stage of the cyber attack life cycle.

- The SOC has a set of standard network and host instrumentation packages that can be adapted to common monitoring scenarios across the constituency.
- The SOC leverages a blend of COTS, FOSS, and custom capabilities in its monitoring and analytic architecture, leveraging the right tool for the right job, as it sees fit.

- The SOC's constituency is instrumented with monitoring packages and data feeds that:
  - Cover the entire cyber attack life cycle
  - Cover the network architecture from edge to core
  - Address threats and attack vectors of relevance to the constituency
  - Target systems and programs that are mission relevant
  - Balance completeness with economy
  - Are adaptable to the environment, architecture, and limitations of end systems
  - Incorporate both network and host sensing and data feeds
  - Mix signature and heuristics-based detection
  - Provide overlapping, complementary observables and techniques, where needed
  - Use a mix of freestanding CND monitoring technologies (e.g., those from Section 8.2. and Section 8.3) with security-relevant data feeds (from Section 9.2).
- The SOC pursues bulk or enterprise licensing to enable economies of scale with its most often-used monitoring packages; for coordinating or tiered SOCs, it provides enterprise licensing for its subordinate SOCs where there is demand for specific tools or technologies.
- The SOC is able to articulate the value it derives from each of its sensors or data feeds, such as through their proportional support to finding an initial incident tip-off, running incidents to ground, or both.
- All monitoring capabilities receive regular attention of analysts and analytic tools; no investments are just "collecting dust."
- All new constituency IT programs, projects, and system owners are compelled through process or mandate to seek the SOC's assistance in applying SOC monitoring capabilities.
- New constituency IT programs, projects, and system owners proactively seek the SOC's assistance in applying CND monitoring capabilities to their systems.
- The SOC has sufficient budgetary and engineering resources to provide CND monitoring capabilities and data collection to programs and projects that request it.
- Other than those owned and operated by the SOC, there are no or very few "rogue" and "one-off" IDS or other CND monitoring systems operated within the constituency.
- The primary drivers for new CND technology investments are the SOC's own:
  - Assessments of gaps in threat detection
  - Knowledge of COTS and FOSS tool and technology improvements
  - Knowledge of changes to constituency architecture, configuration, networks, enclaves, and hosts
  - Budgetary and resource limitations.

- Monitoring capabilities send data (logs, PCAP) and exchange command and control data with the SOC's analytic framework through assured, encrypted communication, to the extent that such mechanisms are technically feasible.
- Passive monitoring capabilities are isolated from the networks they monitor, leveraging techniques discussed in Section 10.
- SOC tools, systems, and workstations have little to no trust relationship with general constituency IT systems.
- The SOC has exclusive administrative rights to all of its:
  - Passive detection capabilities, such as its network sensors
  - Analytics engine(s) such as SIEM
  - Event data, case data, and PCAP data storage systems.

G.3 Analytics and Detection

This section describes the qualities of the analytics and detection tools used by the SOC.

- The SOC leverages a unified analytic framework for incident monitoring, triage, and detection, such as with a SIEM, purpose built for such use.
- The SOC has consolidated all of its security-relevant data feeds into one integrated analytic architecture, such as a SIEM.
- No one data feed or sensor technology holds a majority of analysts' focus; at best, a particular data feed or sensor technology may engender a plurality of attention.
- The SOC applies automated correlation and triage to its real-time data feeds, such as with a SIEM.
- The SOC applies data mining and other techniques to examine historical data for evidence of malicious or anomalous activity.
- Content in the analytic framework (e.g., triaging, filters, and correlation) incorporates knowledge about constituency environment, threats, and vulnerabilities.
- All of the SOC's signature-based detection capabilities such IDSes and indicator lists are tuned or updated with new signatures at least once a week.
- The SOC authors custom rules and SIEM content to enhance its monitoring efforts beyond what is provided by vendors or the open source community.
  - Analytics include use cases that incorporate information about constituency mission and are meant to detect compromise of constituency mission elements.
  - Updates to signatures, content, and heuristics are tracked through a CM process.
  - Indicators of compromise collected from open sources and from other SOCs are regularly integrated into custom signatures and analytics.
  - Custom signatures and other analytical tools are fed back to partner SOCs.

- The SOC tests more complex analytics on a test/dev/preproduction/beta system before applying them to production systems.
- Analysis and monitoring systems are properly tuned so that the majority of analysts' time is focused on interacting and comprehending data, and the minority is spent wading through the system interface or waiting for query results to return.
- The SOC's main data repositories (SIEM, PCAP, malware, and case data) and analysis workstations are located within a protected enclave, as described in Section 10.

G.4 Monitoring

This section describes the qualities of the monitoring tools and processes used by the SOC.

- Analysis functions are split into at least two distinct tiers, including:
  - Tier 1. Responsible for monitoring real-time data feeds and escalating potential cases to Tier 2
  - Tier 2. Responsible for in-depth analysis and response
  - A third analysis tier, if mission ops tempo demands it.
- SOC real-time monitoring operations (such as Tier 1) cover the times of the day and days of the week during which constituency IT systems are being used (likely meaning the SOC provides 12x5, 12x7, or 24x7 coverage).
- Tier 1 analysts are given a set of well-defined views into security-relevant data collected by the SOC:
  - Analysts are able to acknowledge or follow up on all of events that are displayed during their shift.
  - Analysts are neither overwhelmed with alerts nor lacking data to analyze during the course of their shift.
  - Views into data do not comprise unfiltered, raw event feeds.
  - There is no arbitrary "hunting and pecking" for alerts.
  - Use cases are driven by input from each section of the SOC, including Tier 1, and are managed by a designated content or signature manager.
  - Tier 1 use cases are clearly divided among multiple Tier 1 watch standers, and there is no "free for all" in deciding who analyzes what.
- There is regular rotation between use cases among Tier 1 analysts, thereby reducing monotony and repetition, from day to day.
- Tier 1 is given a discrete time threshold during which each scrolling event or other data visualization must be evaluated; if it takes longer than the threshold to run an event to ground, it is escalated to Tier 2.
- Steps for handling the most clear-cut incidents, such as malware infections, are followed by SOC analysts using an up-to-date set of SOPs.

- Analysts are provided workflow and integration between several SOC tools, allowing them to pivot between them; these tools cover:
  - Real-time alert monitoring, triage, correlation, and event drill-down
  - Historical event query and data mining
  - Network traffic analysis and reconstruction
  - Runtime malware execution and detonation
  - Incident ticketing and tracking
  - Indicator and adversary campaign tracking.

G.5 Threat Assessment

This section describes how the SOC understands the adversary and effectively responds to incidents.

- Analysts capture information about adversaries of interest, such as their TTPs, through a knowledge base (such as a database or semantic wiki) that:
  - All analysts have access to
  - Is used by SOC analysts on a daily basis
  - Sits in addition to, but not replacing, a case management system
  - Can be edited by all SOC analysts in an effort to record and learn about adversaries' actions over time
  - Contains a comprehensive list of Indicators of Compromise that the SOC uses and updates.
- The SOC maintains a repository of case evidence such as malware samples, traffic captures, system logs, screenshots, memory images, or disk images:
  - Used to perform trending on adversary TTPs over time
  - Tied to the aforementioned actor/asset knowledge base.
- The SOC exercises the ability to observe the adversary's behavior and determine the true nature and extent of incidents, without being driven to respond with certain countermeasures.
- The SOC has the means to gauge the extent of damage, attackers' motives, attack vector, and probable attacker attribution.
- The SOC has tools supporting rapid runtime analysis of malware behavior, such as a content detonation system.
- The SOC has the expertise, tools, and resources to reverse engineer malware.
- The SOC has the expertise, tools, and resources to perform in-depth digital forensic analysis of media relevant to major intrusions, such as hard drive forensic analysis.
- The SOC is able to fully reconstruct adversary activity in the event of successful intrusions.

- The SOC is able to disseminate a damage assessment to appropriate stakeholders regarding likely impacts to constituency missions relating to intrusion activity, especially data exfiltration or manipulation. The SOC has policy, procedure, tools, and expertise in place to collect, retain, and make copies of digital evidence.
- The SOC has demonstrated the consistent ability to maintain the integrity, chain of custody, and legal admissibility of digital evidence, as applicable to the SOC's legal jurisdiction.
- The SOC has the means to rapidly scan most managed constituency assets for evidence indicating the current or historical presence of an adversary, such as a remote incident response tool set described in Section 8.3.9.

G.6 Escalation, Response, and Reporting

This section describes how incidents are escalated within and outside the SOC, how they are responded to, and how they are reported.

- The SOC has documentation, such as CONOPS or SOPs, that clearly articulates the types of incidents it handles, the manner in which it responds to incidents, and the responsibilities it and its partner organizations have in the incident response process.
- The SOC follows its CONOPS and updates it at least every 36 months.
- The SOC has internal escalation and response SOPs for the incidents it deals with most commonly (malware infections, data spills/leaks, and phishing attacks).
- On average, the SOC is able to contact parties involved in a suspect incident within minutes of needing to do so.
- When the SOC calls a constituent (such as a TA, sysadmin, or user), its position as the coordinating authority for cyber incidents is not disputed, and those parties work with the SOC to carry out incident prevention and response steps as they mutually deem appropriate, and in a timely manner.
- When an incident with a particularly high severity develops, SOC personnel respond in an orderly, calm, and calculated manner.
- When coordinating an incident with a widespread, severe impact, the SOC is able to gather the appropriate constituency executives and involved parties into a meeting within four business hours after initiating contact.
- In response to a confirmed incident, the SOC's detection and response activities are able to identify and expel all adversary footholds within an intrusion, including any later stage malware, account credentials, and lateral movement, with high assurance that response activities were fully successful and that no adversary footholds remain.
- The speed of the SOC's detection and response activities is sufficient to comprehensively expel all adversary footholds before adversaries are able to fulfill their intent,

such as lateral movement, harvesting of account credentials, privilege escalation, internal network mapping, or exfiltration of sensitive data.

- The SOC serves as the distribution point for routine countermeasure directives (e.g., firewall blocks, DNS black holes, or IDS signature deployments).
- The SOC escalates major cases to cognizant organizations internal or external to the constituency (e.g., law enforcement or legal counsel) as needed.
- Lessons learned from notable incidents are fed back to the entire SOC.
- All sections of the SOC use a common system to track incidents from cradle to grave. (See Section 12.1.)
  - It is specifically written or customized to support security incident case tracking (as opposed to general IT).
  - The SOC's case data is not comingled or accessible by non-SOC parties, unless such access has specifically been granted to said parties.
  - If applicable, it supports or conforms to relevant external incident reporting requirements that the SOC may be subject to.
  - Is used by SOC managers to ensure quality, timeliness, and correctness of incident analysis, escalation, response, and closure.

G.7 Situational Awareness

This section covers the SOC's own cyber SA and how it provides that awareness to external parties.

- For non-coordinating/non-tiered SOCs, at least one person in the SOC is able to describe the following qualities of each Internet gateway, major data center, large campus, and major project/system:
  - The structure of its networks and external connections
  - The general types of computing assets deployed on them
  - General sense of attack surface and stand-out vulnerabilities
  - Likely threats against those systems
  - Supported mission or business functions.
- For non-coordinating SOCs, the SOC gathers network-mapping data through active device enumeration and interrogation, collection and analysis of network device configurations, manually gathering network maps, or a combination thereof, covering each area of the constituency at least every 12 months.
- The SOC indoctrinates new personnel in details of constituency networks, assets, and mission, thereby ensuring deep knowledge spreads from veterans to junior analysts.
- For each major adversary that has successfully attacked the constituency in the last 24 months, or is considered to be of substantial concern, at least one person in the SOC

is able to articulate the following qualities of that adversary, including the confidence and gaps they have in such knowledge:
- Capability, including skill level and resources
- Intent and motivation
- Probability of attack
- Level of access (legitimate or otherwise)
- Impact to constituency business/mission and IT
- Likely identity or allegiance
- Actions: in the past, present, and projected into the future.

- The SOC routinely consumes cyber threat intelligence from a wide variety of sources, including but not limited to:
  - Open source
  - Vendor subscriptions
  - Independent security researchers
  - Community cyber threat-sharing forums
  - Bilateral and multilateral sharing agreements with peers.

- The SOC actively reviews outside cyber intelligence reporting and cyber intelligence feeds to ensure they are of high fidelity prior to acceptance into the SOC's cyber threat intelligence repositories and sensors.

- The SOC has a process to routinely update high-fidelity watch lists, block lists, reputation filters, and other means of prevention and detection into its broader security infrastructure for automated use.

- The SOC actively disseminates actionable cyber threat indicators and intelligence, derived from its own observations and analytical results, to community threat-sharing partners within its SOC peer group (such as education, industry, commerce, government, not for profit, etc.) on at least a bimonthly basis.

- The SOC synthesizes SA information for the constituency, such as:
  - Routine weekly cyber news digests
  - Non-routine emergency alerts and warnings
  - Annual or biannual cyber threat assessments
  - Network mapping or vulnerability scan metrics
  - Incident reporting metrics.

- The SOC posts SA information on its website, available to designated members of the constituency.

- The SOC can articulate the limits of its SA in detail, in terms of monitoring coverage, cyber attack life cycle coverage, and threat parity.

G.8 Prevention

This section covers the SOC's impact to the constituency cybersecurity program supporting incident prevention.

- The SOC uses its SA to drive remediation of constituency vulnerabilities and poor security practices.
- The SOC provides consulting on cyber threats, security architecture, and best practices to constituency IT programs and lines of business.
- The SOC participates in formulating (and possibly providing) security "general hygiene" training and education to constituents, such as safe browsing and email tips.
- The SOC is regularly consulted on matters of cybersecurity policy by constituents.
- The SOC is regarded by constituents as an organization that helps constituency mission and business operations, rather than hindering them.
- The SOC has deployed intrusion prevention capabilities such as NIPS, HIPS, and content detonation devices to key points in the enterprise.
- Prevention technologies beyond host-based AV are set to block critical attacks, and actually do so at least on a monthly basis.
- The SOC hosts prevention capabilities (such as AV and HIDS/HIPS programs and signatures) on its website for easy download by constituents.
- The SOC provides clear guidance and means to constituents who wish to submit potential incidents to the SOC, or to seek the SOC's help in other cybersecurity matters.

G.9 Training and Career

This section covers topics related to the care and feeding of SOC personnel, from the time they are interviewed to the time they leave.

- The head of the SOC has ultimate authority regarding who is selected and who is turned down for employment with the SOC.
- The SOC has a consistent means of vetting personnel for employment, to include its own set of qualification standards and interview process.
- SOC personnel receive compensation commensurate with the importance of their positions and geographic adjustments that maintain parity with security operations in industry.
- New SOC personnel go through a consistent indoctrination process that includes an orientation on SOC mission, personnel, capabilities, policies, and SOPs.
- The SOC maintains a technical qualification process for new recruits that vets them for their ability to complete core job functions, through a written exam and/or lab practical.

- All SOC personnel on the job more than three months have passed the technical qualification tests for their particular position.
- At least two people in the SOC have sufficient (if not deep) knowledge of each of the areas of expertise listed in Section 7.1.3.
- At least two weeks of paid training is granted to every SOC analyst every year; this training may be composed of attendance at professional cybersecurity conferences such as those mentioned in Section 7.3.2, and/or technical class-based training on new tools, offensive techniques, or defensive techniques.
- Each SOC team member who has been on the job for at least 24 months has been cross-trained on at least one SOC job function outside his or her normal daily routine.
- All members of the SOC in senior or lead positions are trained in penetration testing and other adversarial techniques.
- SOC leads regularly take opportunities to engage SOC team members in team-building activities inside and outside of the workplace.
- SOC personnel who demonstrate exceptional on-the-job performance are recognized for their efforts and advanced to more senior positions in the SOC, such as intel fusion, engineering, or penetration testing.
- The SOC's annual attrition rate is less than 35 percent.

For another take on measuring SOC maturity, see [290].

Appendix H
Glossary

Advanced persistent threat (APT)	A well-resourced, sophisticated adversary that uses multiple attack vectors such as cyber, physical, and deception to achieve its objectives. An APT pursues its objectives repeatedly over an extended period of time, adapts to the defenders' efforts to resist the APT, and is determined to maintain the required interaction level to execute its objectives [291].
Adversary	An individual, group, organization, or government that conducts or has the intent to conduct detrimental activities [292].
Alert	An event or notification generated by a monitoring system (e.g., an IDS), usually with an assigned priority.
Asset	A major application, general support system, high-impact program, physical plant, mission-critical system, personnel, equipment, or a logically related group of systems [42].
Attack	Any kind of malicious activity that attempts to collect, disrupt, deny, degrade, or destroy information system resources or the information itself [42].
Artifact	The remnants of an intruder attack or incident activity. These could be software used by intruder(s), a collection of tools, malicious code, logs, files, output from tools, or status of a system after an attack or intrusion. Examples of artifacts range from Trojan horse programs and computer viruses to programs that exploit (or check for the existence of) vulnerabilities or objects of unknown type and purpose found on a compromised host [8].
Audit data	The data that comprises a security audit trail written by a host OS or application. (See **Audit trail**.)
Audit log	See **Audit trail**.
Audit review	The process whereby a human analyst uses manual and automated means to inspect the details of a computer security audit trail for evidence of potentially malicious or anomalous activity or other activity of concern.
Audit trail	A chronological record that reconstructs and examines the sequence of activities surrounding or leading to a specific operation, procedure, or event in a security-relevant transaction from inception to final result [42].[1]

---

1   It should be noted that this definition of a security audit trail portrays the ideal case for security logs, not the reality. The reality is that audit logging functions in many programs record only a portion of the information necessary to reconstruct the who, what, when, and where of activity; or that they do so but in an obscured or hard-to-reconstruct fashion.

Blue Team	The group responsible for defending an enterprise's use of information systems by maintaining its security posture against a group of mock attackers (the Red Team). The Blue Team typically must defend against real or simulated attacks (1) over a significant period of time, (2) in a representative operation context such as part of an operational exercise, and (3) according to rules established with the help of a neutral group that is moderating the simulation or exercise [42].
Case	A written record of the details surrounding a potential or confirmed computer security incident.
Chief information officer (CIO)	Senior-level executive responsible for (1) providing advice and other assistance to the head of the executive organization and other senior management personnel of the organization to ensure that information systems are acquired and information resources are managed in a manner that is consistent with laws, directives, policies, regulations, and priorities established by the head of the organization; (2) developing, maintaining, and facilitating the implementation of a sound and integrated information system architecture for the organization; and (3) promoting the effective and efficient design and operation of all major information resources management processes for the organization, including improvements to work proce sses of the organization (adapted from [42]).
Chief information security officer (CISO)	The senior-level executive within an organization, responsible for management of the information security program for the entire organization (usually subordinate to the CIO).
Computer forensics	The practice of gathering, retaining, and analyzing computer-related data for investigative purposes, in a manner that maintains the integrity of the data [42].
Computer network defense (CND)	The practice of defense against unauthorized activity within computer networks, including monitoring, detection, analysis (such as trend and pattern analysis), and response and restoration activities [42].
Computer security incident response team (CSIRT)	See **Security operations center**.
Constituency	The group of users, sites, IT assets, networks, and organizations to which the CSIRT provides services (adapted from [43]).
Correlation	Near-real-time finding of relationships among multiple events from different sources [211].

Cracker	Someone who tries to gain unauthorized access to a host or set of hosts or networks, often with malicious intent (adapted from [44]). (See also **Advanced persistent threat**, **Adversary**, **Script kiddie**.)
Cyberspace	A global domain within the information environment consisting of the inter-dependent network of information systems infrastructures, including the Internet, telecommunications networks, computer systems, and embedded processors and controllers [42].
Cyber attack life cycle	The entire progression of the cyber intrusion, from initial reconnaissance through exploitation and persistence (adapted from [293]).
Cyber intelligence report (cyber intel)	Formal and informal reports from SOCs, commercial vendors, independent security researchers, or independent security research groups that discuss information about attempted or confirmed intrusion activity, threats, vulner-abilities, or adversary TTPs, often including specific attack indicators.
Cyber kill chain	See **Cyber attack life cycle**.
Cyber observable	Events or stateful properties that can be seen in the cyber operational domain (adapted from [294]).
Cybersecurity	Measures to provide information assurance, improve resilience to cyber inci-dents, and reduce cyber threats (adapted from [295]).
Cybersecurity operations center	See **Security operations center**.
Cyber situational awareness	Within a volume of time and space, the perception of an enterprise's secu-rity posture and its threat environment, the comprehension/meaning of both taken together (risk), and the projection of their status into the near future [42].
Enclave	A collection of information systems connected by one or more internal net-works under the control of a single authority and security policy. The sys-tems may be structured by physical proximity or by function, independent of location [42].
Event	Any observable occurrence in a system and/or network. Events can indicate that an incident is occurring [42].
Exfiltrate (exfil)	The act of sending sensitive information out of a network enclave against established security policy or controls, usually in a surreptitious manner.
False positive	A circumstance under which a monitoring system generated an alert, but no malicious activity actually occurred.
Host	A computer or other network-enabled device such as a smartphone or net-work-enabled printer.

Host intrusion detection system (HIDS)	An intrusion detection system (IDS) that operates on information collected from within an individual computer system. This vantage point allows host-based IDSes to determine exactly which processes and user accounts are involved in a particular attack on the OS. Furthermore, unlike network-based IDSes, host-based IDSes can more readily "see" the intended outcome of an attempted attack because they can directly access and monitor the data files and system processes usually targeted by attacks [42].
Incident	An assessed occurrence that actually or potentially jeopardizes the confidentiality, integrity, or availability of an information system or the information the system processes, stores, or transmits; or that constitutes a violation or imminent threat of violation of security policies, security procedures, or acceptable use policies [42].
Indicator	A recognized action (specific, generalized, or theoretical) that an adversary might be expected to take in preparation for an attack [42].
Information assurance (IA)	Measures that protect and defend information and information systems by ensuring their availability, integrity, authentication, confidentiality, and non-repudiation. These measures include providing for restoration of information systems by incorporating protection, detection, and reaction capabilities [42].
Information system security manager (ISSM)	The individual responsible for the information assurance of a program, organization, system, or enclave [42].
Information systems security officer (ISSO)	The individual assigned responsibility for maintaining the appropriate operational security posture for an information system or program [42].
Insider threat	An entity with authorized access (i.e., within the security domain) that has the potential to harm an information system or enterprise through destruction, disclosure, modification of data, and/or denial of service [42].
Inspector General (IG)	A government organization whose primary responsibilities are to detect and prevent fraud, waste, abuse, and violations of law and to promote economy, efficiency, and effectiveness in the operations of the federal government [296].
Intrusion	Unauthorized act of bypassing the security mechanisms of a system [42].
Intrusion detection system (IDS)	Hardware or software products that gather and analyze information from various areas within a computer or a network to identify possible security breaches, which include both intrusions (attacks from outside the organizations) and misuse (attacks from with the organizations) [42].
Intrusion prevention system (IPS)	A system that can detect an intrusive activity and can also attempt to stop the activity, ideally, before it reaches its targets [42].

Log	See **Security audit trail.**
Malware	See **Malicious code.**
Malicious code	Software or firmware intended to perform an unauthorized process that will have adverse impact on the confidentiality, integrity, or availability of an information system. A virus, worm, Trojan horse, or other code-based entity that infects a host. Spyware and some forms of adware are also examples of malicious code [42].
NetFlow	A network protocol developed by Cisco Systems that provides network administrators with access to IP flow information from their data networks. A flow is a unidirectional sequence of packets with some common properties that pass through a network device [139].
NetFlow data	Data exported as a result of the NetFlow service, including IP addresses, packet and byte counts, time stamps, type of service, application ports, and input and output interfaces [139].
Network	Information system(s) implemented with a collection of interconnected components including routers, hubs, cabling, telecommunications controllers, key distribution centers, and technical control devices [42].
Network intrusion detection system (NIDS)	IDSes that detect attacks by capturing and analyzing network packets. Listening on a network segment or switch, one network-based IDS can monitor the network traffic affecting multiple hosts that are connected to the network segment [42].
Ops tempo	A relative measure of the pace of an operation in terms of frequency and rhythm of activities and duties performed.
Packet capture (PCAP)	Network traffic recorded in a libpcap-formatted file.
Penetration testing	A test methodology in which assessors (typically working under specific constraints) attempt to circumvent or defeat the security features of an information system [42].
Red Team	A group of people authorized to emulate a potential adversary's attack or exploitation capabilities against an enterprise's security posture. The Red Team's objective is to improve enterprise IA by demonstrating the impacts of successful attacks and by demonstrating what works for the defenders in an operational environment. (See also **Blue Team** [42].)
Remote access tool	Programs that run on a remote host, providing an intruder with control over the target system; often used by adversaries to maintain remote persistence without the user's knowledge or consent.

Script kiddie	A would–be attacker who has a relatively low skill level, such as someone who can download and execute scripts or other exploits, but is unable to create their own attacks; a pejorative term usually in reference to a novice cracker (adapted from [44]).
Security operations center (SOC)	A team composed primarily of security analysts organized to detect, analyze, respond to, report on, and prevent cybersecurity incidents (adapted from [42] and [43]).
Security information and event management (SIEM)	A system designed to collect, aggregate, filter, store, correlate, triage, and display various security-relevant data feeds.
Situational awareness (SA)	See **Cyber situational awareness**.
System administrator (sysadmin)	A person who is responsible for day–to–day technical duties of configuring, maintaining, and administering an IT asset or collection of assets (adapted from [44]).
System owner	A person or organization who is recognized as the business owner of an IT asset.
Tactics, techniques, and procedures (TTPs)	The employment and ordered arrangement of forces in relation to each other (tactics); nonprescriptive ways or methods used to perform missions, functions, or tasks (techniques); and standard, detailed steps that prescribe how to perform specific tasks (procedures) (from [297]); in a cyber context, used to refer to the entirety of how an attacker or defender operates.
Tier 1 CND analysts	Members of the SOC who are responsible for real-time monitoring of sensor and data feeds, triage of events of interest, and escalation to Tier 2.
Tier 2 CND analysts	Members of the SOC who are responsible for collection and in–depth analysis of a wide variety of indicators and log data and coordination of incident response activities.
Threat	Any circumstance or event that has the potential to adversely impact organizational operations (including mission, functions, image, or reputation), organizational assets, individuals, other organizations, or the nation, through an information system via unauthorized access, destruction, disclosure, modification of information, and/or denial of service [42].
Triage	The process of receiving, initial sorting, and prioritizing of information to facilitate its appropriate handling [8].
Virus	A computer program that can copy itself and infect a computer without permission or knowledge of the user. A virus might corrupt or delete data on a computer, use email programs to spread itself to other computers, or even erase everything on a hard disk [42].

Vulnerability	A weakness in an information system, system security procedures, internal controls, or implementation that could be exploited by a threat source [42].
Vulnerability assessment	The systematic examination of an information system or product to determine the adequacy of security measures, identify security deficiencies, and provide data from which to predict the effectiveness of proposed security measures and confirm the adequacy of such measures after implementation [42].

Appendix I
List of Abbreviations

ACL	Access Control List
AES	Advanced Encryption Standard
APT	Advanced Persistent Threat
ASCII	American Standard Code for Information Interchange
ASIC	Application Specific Integrated Circuit
ATM	Asynchronous Transfer Mode
AV	Antivirus
B2B	Business to Business
B2G	Business to Government
BIOS	Basic Input/Output System
BSD	Berkley Software Distribution
BSOD	Blue Screen of Death
CBE	Common Base Event
CEE	Common Event Expression
CEF	Common Event Format
CEI	Common Event Infrastructure
CEO	Chief Executive Officer
CERT	Computer Emergency Readiness Team
CI	Counterintelligence
CIDF	Common Intrusion Detection Framework
CIFS	Common Internet File System
CIO	Chief Information Officer
CIRC	Computer Incident Response Center
CIRT	Computer Incident Response Team
CISO	Chief Information Security Officers
CSIRT	Computer Security Incident Response Team
CJCSI	Chairman of the Joint Chiefs of Staff Instruction
CM	Configuration Management
CMMI	Capability Maturity Model Integration
CNA	Computer Network Attack
CND	Computer Network Defense

CNE	Computer Network Exploitation
CNSS	Committee on National Security Systems
CONOPS	Concept of Operations
COO	Chief Operating Officer
COOP	Continuity of Operation
COTS	Commercial Off-the-Shelf
CPU	Central Processing Unit
CRITs	Collaborative Research into Threats
CS	Computer Science
CSIRC	Computer Security Incident Response Center
CSIRT	Computer Security Incident Response Team
CSO	Chief Security Officer
CSOC	Cybersecurity Operations Center
CSV	Comma-Separated Values
CTAC	Cyber Threat Analysis Cell
CTO	Chief Technical Officer
CUDA	Compute Unified Device Architecture
CVE	Common Vulnerabilities and Exposures
CybOX	Cyber Observables eXpression
DASD	Direct-Attached Storage Device
DHCP	Dynamic Host Configuration Protocol
DHS	Department of Homeland Security
DLP	Data Loss Prevention
DMVPN	Dynamic Multipoint Virtual Private Network
DMZ	Demilitarized Zone
DoD	Department of Defense
DoJ	Department of Justice
DoS	Denial of Service
FAQ	Frequently Asked Question
FBI	Federal Bureau of Investigation
FIRST	Forum for Incident Response and Security Teams
FOC	Full Operating Capability
FOSS	Free or Open Source Software
FPGA	Field Programmable Gate Array

FTE	Full Time Equivalent
FTP	File Transfer Protocol
G2G	Government to Government
GB	Gigabyte
GPGPU	General Purpose Graphics Processing Unit
GPO	Group Policy Object
gigE	Gigabit Ethernet
GRE	Generic Routing Encapsulation
GUI	Graphical User Interface
HIDS	Host Intrusion Detection System
HIPS	Host Intrusion Prevention System
HOPE	Hackers on Planet Earth
HR	Human Resources
HTTP	Hypertext Transfer Protocol
IaaS	Infrastructure as a Service
I/O	Input/Output
IA	Information Assurance
ICMP	Internet Control Message Protocol
ID	Identity; Identification
IDPS	Intrusion Detection and Prevention Systems
IDS	Intrusion Detection System
IG	Inspector General
IODEF	Incident Object Description and Exchange Format
IP	Internet Protocol
IPS	Intrusion Prevention System
ISAC	Information Sharing and Analysis Center
ISSE	Information Systems Security Engineer
ISSM	Information System Security Manager
ISSO	Information Systems Security Officer
IT	Information Technology
JDBC	Java Database Connectivity
KVM	Keyboard, Video, Mouse
LAN	Local Area Network
LDAP	Lightweight Directory Access Protocol

LE	Law Enforcement
LM	Log Management
MAC	Media Access Control
Mb	Megabit
MD5	Message-Digest Algorithm (5)
MOA	Memorandum of Agreement
MOU	Memorandum of Understanding
MPLS	Multiprotocol Label Switching
MSSP	Managed Security Service Provider
NAC	Network Access Control
NAS	Network Area Storage
NAT	Network Address Translation
NICS	Network Interface Cards
NIDS	Network Intrusion Detection System
NIPS	Network Intrusion Prevention System
NIS	Network Information Service
NIST	National Institute of Standards and Technology
NMS	Network Management System
NOC	Network Operations Center
NOSC	Network Operations and Security Center
O&M	Operation and Maintenance
OODA	Orient, Observe, Decide, and Act
OS	Operating System
OSI	Open Source Interconnection
OSSIM	Open Source Security Information Management
PaaS	Platform as a Service
PCAP	Packet Capture
PCIe	Peripheral Component Interconnect express
PDF	Portable Document Format
PE	Portable Executable
PF	Packet Filter
PID	Process Identification Number
POC	Point of Contact
PoP	Point of Presence

POP3	Post Office Protocol 3
R&D	Research and Development
RAID	Redundant Array of Independent Disks
RAM	Random Access Memory
RAT	Remote Access Tool
RDBMS	Relational Database Management System
RSA	Rivest, Shamir, and Adleman (public key cryptosystem)
RT	Request Tracker
RTF	Rich Text Format
RX	Receiver
SA	Situational Awareness
SaaS	Software as a Service
SAN	Storage Area Network
SCADA	Supervisory Control and Data Acquisition
SDEE	Security Device Event Exchange
SDK	Software Development Kit
SEI	Software Engineering Institute
SHA	Secure Hash Algorithm
SIEM	Security Information and Event Management
SLA	Service Level Agreement
SMB	Server Message Block
SMTP	Simple Mail Transfer Protocol
SNMP	Simple Network Management Protocol
SOC	Security Operations Center
SONET	Synchronous Optical Network
SOP	Standard Operating Procedure
SPAN	Switched Port Analyzer
SQL	Structured Query Language
SSD	Solid State Drive
SSH	Secure Shell
SSL	Secure Socket Layer
SSO	Single Sign-On
STE	Secure Terminal Equipment
STIX	Structured Threat Information eXpression

STU	Secure Telephone Unit
TA	Trusted Agent
TACACS	Terminal Access Controller Access Control System
TAXII	Trusted Automated eXchange of Indicator Information
TB	Terabyte
TCO	Total Cost of Ownership
TCP	Transmission Control Protocol
TLS	Transport Layer Security
TPM	Trusted Platform Module
TTPs	Tactics, Techniques, and Procedures
TX	Transmit
UDP	User Datagram Protocol
URL	Uniform Resource Locator
USB	Universal Serial Bus
VA/PT	Vulnerability Assessment/Penetration Testing
VIP	Very Important Person
VLAN	Virtual Local Area Network
VM	Virtual Machine
VoIP	Voice over Internet Protocol
VPN	Virtual Private Network
VRF	Virtual Routing and Forwarding
WAN	Wide Area Network
WELF	WebTrends Enhanced Log File
WINE	Worldwide Intelligence Network Environment
WOMBAT	Worldwide Observatory of Malicious Behaviors and Attack Threats
XML	Extensible Markup Language
ZFS	Z File System

Index

A

advanced persistent threat (APT) 4–5,
40–43, 221–223, 319
agility 40–43, 49–70
analysis
  forensic 12, 21
  incident 20, 83
  remote 61–62
analysts 87–107, 317
  background 89–90
  career progression 102–104
  resources 301–306
  skill set 90–92
  turnover 100–107
anti-virus (AV) tools 145–146
application blacklisting 147–148
application whitelisting 148
audit 22, 38–39, 85, 280, 319
authority 71–79
  inherited 75
  written 71–75

C

call center 19, 83
centralized SOCs 60–64
  small and large 60–61
  with Follow the Sun 65
computer network defense (CND) 3
  consolidated 44–48
  definition 8
  team 8
configuration tracking 148–149
continuity of operations (COOP) 62–64

cyber attack life cycle 30–32, 321
cyber intel, cyber intelligence (intel) 13
  analysis and trending 19, 96
  definition 224
  where to get it 244–249
cybersecurity operations center (CSOC).
See security operations center (SOC)
cyber situational awareness (SA) 24, 25–28
  areas of 26–28
  ways to gain 26
cyber threat analysis cell (CTAC) 221–244
  definition 222–223
  deliverables 224–226
  integration 224, 225–230
  investments needed 231–240
  prerequisites 230–231
  space requirements 240
  staffing 238–240
  standing up a 240–243

D

data feeds
  leveraging for audit 204–205
  maintenance 203
  obtaining 201–203
  tuning 193–201
  types 191–193
data quality 32
data quantity 38–40
data sources
  comparisons 194–200
  selecting and instrumenting 188–207

E

event
  definition 10

F

false positives 34, 36–38, 123, 321
firewalls 150
full session capture 130–132

H

honeypots 141
host intrusion prevention system
(HIPS) 12, 146–147
host monitoring and defense 142–154
host sensors 181–182

I

incident
  definition 11, 322
  prevention 317
  response 20, 250–254, 314–315
  tip-offs 29–34, 38
  tracking 254–257
information technology
  enterprise 5
information technology (IT) 3
  enterprise 3
insider threat 322
intrusion detection 35–36, 118–122, 141,
146–147, 322

M

malware 323
  detonation and analysis 138–141
  reverse engineering 12

N

NetFlow 108, 126–129, 323
network access control (NAC)
systems 149–150
network hub 207–208, 210

network intrusion prevention system
(NIPS) 12
network mapping 23, 111–114
network monitoring 118–142
  platforms 132–138
network sensors
  active 180
  isolating 207–212
  passive 176–180
network tap 208–210

O

OODA (orient, observe, decide, and act)
Loop 26, 40

P

passive fingerprinting 116
policies
  for monitoring coverage 186–187
  IT and cybersecurity 74–75

S

security information and event
management (SIEM) 12, 109, 154–175, 324
security operations center (SOC) 3
  agility 40–43, 49–70
  authority 15, 17–18
  capabilities of 18
  characteristics of 15–18, 307–318
  constituency 15, 50–53
  coordinating 69–70
  definition 9–10, 324
  document library 295–300
  enclave design 212–218
  engineering staff 99–100
  evaluating the need for 282–285
  external interfaces 278–281

hours of operation 291–294
importance of collaboration 243
IT operations 78–79
large 56–58
mission 10, 206–220
organizational models 15–16, 50–56
organizational placement 75–78
outsourcing 283–285
physical location 58–70
roles and incident escalation 13–14
service templates 81–85, 82–83
size vs. agility 49–70
small 54–56
staffing 87–107, 317–318
standing up a new 286–290
structuring 53–57
templates 51–53
tiered 66–69
what it is not 13–14
sensor
  cost 184–186
  placement 175–188
  tuning and maintenance 21, 84, 98–99
sharing sensitive information 218–220
situational awareness (SA) 315–316, 324.
See cyber situational awareness (SA)
switched port analyzer (SPAN) 207–210

T

tactics, techniques, and procedures
(TTPs) 13, 85, 220–221, 324
The MITRE Corporation 3
threat assessment 313–314
Tier 1 11–12, 94–95, 227, 229, 324
Tier 2 11–13, 95–96, 227, 229, 305–306, 324
tip-offs 29–34, 38, 192
traffic metadata 129–130

traffic redirection 207–211

U

user activity monitoring 150

V

virtualization 187–188
vulnerability 325
  assessment 23, 84, 97–98
  scanning 23, 84, 97, 115–116

Ten Strategies of a World-Class Cybersecurity Operations Center conveys MITRE's accumulated expertise on enterprise-grade computer network defense. It covers ten key qualities of leading Cybersecurity Operations Centers (CSOCs), ranging from their structure and organization, to processes that best enable effective and efficient operations, to approaches that extract maximum value from CSOC technology investments. This book offers perspective and context for key decision points in structuring a CSOC and shows how to:

The MITRE Corporation is a not-for-profit organization that operates federally funded research and development centers (FFRDCs). FFRDCs are unique organizations that assist the U.S. government with scientific research and analysis, development and acquisition, and systems engineering and integration. We're proud to have served the public interest for more than 50 years.

- Find the right size and structure for the CSOC team
- Achieve effective placement within a larger organization that enables CSOC operations
- Attract, retain, and grow the right staff and skills
- Prepare the CSOC team, technologies, and processes for agile, threat-based response
- Architect for large-scale data collection and analysis with a limited budget
- Prioritize sensor placement and data feed choices across enterprise systems, enclaves, networks, and perimeters

If you manage, work in, or are standing up a CSOC, this book is for you. It is also available on MITRE's website, www.mitre.org.

The MITRE Corporation
202 Burlington Road
Bedford, MA 01730-1420
(781) 271-2000

7515 Colshire Drive
McLean, VA 22102-7539
(703) 983-6000
www.mitre.org

Carson Zimmerman is a Principal Cybersecurity Engineer with The MITRE Corporation. He has ten years of experience working with various CSOCs to better defend against the adversary. He has held roles in the CSOC ranging from tier 1 analyst to architect.

$0.00
ISBN 978-0-692-24310-7
90000>

9 780692 243107

MITRE